BEYOND RECYCLING

BEYOND RECYCLING

A RE-USER'S GUIDE

336 Practical Tips
Save Money and
Protect the Environment

Kathy Stein

Clear Light Publishers
Santa Fe, New Mexico

Clear Light Publishers, 823 Don Diego, Santa Fe, New Mexico 87501

Library of Congress Cataloging-in-Publication Data

Stein, Kathy, 1949– .
Beyond recycling: a re-user's guide: 336 practical tips: save money and protect the environment / Kathy Stein.
p. cm.
Includes index.
ISBN: 0-940666-92-8 (pbk.)
1. Salvage (Waste, etc.) 2. Recycled products. I. Title.
TP995.S84 1997 97–4301
628.4'458—dc21 CIP

DISCLAIMER: Neither Clear Light Publishers nor the author is responsible for any damages, losses, or injuries incurred as a result of any of the material contained in this book.

In preparing this book, the author and publisher have endeavored to describe only practices that are both legal and reasonably safe, and have tried to point out significant hazards. However, readers using ideas in this book must rely on their own personal judgment and take reasonable precautions.

Mention of companies, trade names, products, or services should not be interpreted as conveying approval, endorsement, or recommendation.

FIRST EDITION
10 9 8 7 6 5 4 3 2

Printed in Canada

recycled paper

The paper for text and end sheets used in this publication meets the minimum requirements of American National Standard for Information Sciences—Permanence of Paper for Printed Library Materials, ANSI Z39, 48-1984.

Typography Design/Layout: Vicki S. Elliott

The publisher would like to acknowledge the contributions made by Lt. Col. Alan Robbins, Ret. (RAC) as a leader and pioneer recycler.

This book is
dedicated to the next
seven generations

CONTENTS

APPENDICES

ACKNOWLEDGMENTS

I would like to give special thanks to the many people of good will who graciously shared with me the valuable information that has made this book a reality; to my agent, Susan Travis, for her faith in this project; to my publishers, Harmon Houghton and Marcia Keegan, whose enthusiasm and moral support allowed me to follow my vision and find my voice; to my editors, Sara Held and Ellen Goldberg, for their skillful suggestions and scrupulous attention to detail. Above all, I am deeply grateful to my husband, John, for his steadfast support and many well-timed words of encouragement.

INTRODUCTION

Good News—Congratulations, America! Your recycling efforts are paying off. Thanks to widespread citizen participation in recycling collection programs, 24 percent of our nation's municipal waste stream was either recycled or composted in 1994.

Although we're off to a great start, we still have a terrific challenge before us. In spite of our recycling efforts, the Great American Waste Stream continues to swell. According to the U.S. Environmental Protection Agency (EPA), the total amount of municipal trash we generate annually as a nation is expected to grow to 223 million tons by the year 2000. Clearly, recycling alone is not enough. We will need to find additional ways of reducing the amount of garbage we create.

Beyond Recycling—As recycling programs proliferate around the country, attention is beginning to shift to another method of waste management which not only conserves more energy and natural resources than recycling, but also costs less than recycling. The "new" method is really quite old-fashioned; it is known as re-use.

What Is Re-use?—Re-use is the use, in the same form it was produced, of a material or product that might otherwise be discarded. Returning a soda or beer bottle to be refilled is a familiar example of re-use. Replacing disposable paper towels, napkins, and diapers with washable cloth counterparts are other ways to re-use.

Re-use Has Many Faces—The possibilities for re-use are virtually endless. For example:

- Buying a used computer instead of a new one;
- Attentively maintaining your car so it lasts longer;
- Taking your own re-usable canvas bag to the grocery store;
- Reupholstering a raggedy old sofa instead of throwing it out and buying a new one;

- Replacing disposable alkaline batteries with rechargeable nicad batteries;
- Repairing durable goods such as toasters and washing machines instead of dumping them in the trash when they fail;
- Donating outgrown clothing and still-useful household items to charitable groups that distribute them to individuals who need and can use them.

Re-use Is Energy Efficient—Re-using a product again in its already-manufactured form requires far less energy than recycling. Washing and sterilizing a bottle so it can be refilled, for example, requires significantly less energy than crushing it, melting it down, and manufacturing a new "recycled" one. By remanufacturing laser toner cartridges and by reconditioning household appliances, it's possible to conserve as much as 85 percent of the energy needed to make such products in the first place.

Energy Savings Benefit the Planet in Far-Reaching Ways—Using less energy often means burning less coal or oil, which results in fewer emissions of carbon dioxide, the primary greenhouse gas. In the case of coal-derived energy, less energy used also means less acid rain produced, less air pollution, and less strip mining.

Re-use Reduces Environmental Damage—By conserving natural resources that would otherwise be needed to manufacture replacement products, re-use reduces environmental damage caused by mining and lumbering operations. Many open pit mines, for example, leak acid and heavy metals into surface- and groundwater, virtually killing thousands of miles of rivers. And extensive timbering often causes soil erosion, increases flooding, and destroys wildlife habitat.

Re-use Is Economical—What's good for the environment is often good for your pocketbook too. Over time, for example, washable cotton kitchen towels cost significantly less than single-use paper towels. Rebuilt and recharged laser toner cartridges perform as well as—or better than—new ones, but cost about 40 percent less.

Re-use can even shrink your tax bill. When you give used items to a nonprofit organization, you may be able to claim the fair market value of donated goods as a charitable contribution on your next tax return.

Re-use Creates Jobs for Americans—Businesses that repair and rebuild products such as auto parts, household appliances, computers, and office machines provide jobs for local skilled laborers. Most laser toner cartridge rechargers, for instance, are local American businesses, while the majority of new toner cartridges are made in Asia.

Re-use Is Gaining Ground—Traditional approaches to channeling old goods to new users, such as thrift shops and used car dealers, are increasingly being supplemented by innovative new outlets for preowned goods. At an expanding network of upscale resale stores and consignment boutiques, America's bargain hunters are snapping up secondhand clothes, computers, sporting goods, toys, video games, compact discs, furniture and even musical instruments. As of this writing, **Funcoland®** reports that it has established 173 used video game stores in the last eight years. And **Play It Again Sports®,** launched in 1983, now boasts 695 used sporting goods franchises in the United States and Canada.

Profitable re-use oriented businesses are springing up across the country. In North America alone there are now nearly 200 used building materials stores and more than 8,000 laser toner cartridge recharging companies. In cities across the nation, discarded shipping pallets that once went to landfill are being reclaimed by pallet refurbishers who repair and resell them. The result? A booming new industry that sells an estimated $3 billion of remanufactured pallets annually.

In the nonprofit sector, a burgeoning number of Teacher Resource Centers gather clean, safe industrial discards from local area businesses; then make them available to budget-stressed educators, who use them for classroom crafts and other creative hands-on learning projects.

Racquets for Kids, a program managed by the Tennis Industry Association, collects serviceable old racquets and distributes them to recreation programs and young players who cannot afford to purchase them.

In yet another waste-reducing experiment, a growing number of communities now stage re-use events, special occasions organized to provide residents with an opportunity to give away, swap, or sell re-usable discards. Re-use events from Vermont to California offer residents an opportunity to unload—or acquire—everything from paint to refrigerators.

Big business is discovering that re-use is good for the bottom line. Xerox Corporation's profitable remanufacturing operations reclaim and remanufacture in excess of $200 million worth of piece parts annually. What's more, the company's design standards now mandate that all future products be designed to maximize parts remanufacture.

Along with other manufacturers, Xerox has been slashing disposal costs and boosting profits by insisting that suppliers utilize re-usable shipping containers. Xerox expects its re-usable shipping container program to save up to $15 million per year while reducing waste by 10,000 tons annually.

Just ten months after establishing an office supply depot, where employees can exchange surplus office supplies, AT&T Paradyne Corporation reports that workers have saved thousands of dollars by re-using binders, file folders, hanging folders, **Rolodex®** cards, pens, pencils, dividers, and binder clips.

Even governments are re-using. Many states now sponsor waste exchanges. Operating on the principle that one company's waste may be of value to another company, waste exchanges maintain databases of industrial scraps and chemical leftovers for sale or trade. Thus metal drums discarded by one business can furnish essential raw materials to a company that refurbishes metal drums. Or a pharmaceutical company's acetone waste can be used by another company to make fiberglass for yachts. It has been estimated that in 1990 U.S. waste exchanges saved the nation's industries $27 million by reducing disposal costs for waste producers and raw materials costs for waste re-users.

Re-use Has a Healing Impact on the Earth—Re-use is for people who want to be part of the solution. Americans who choose to re-use deserve a pat on the back for:

- Dumping fewer pollutants into the air and fostering healthier human lungs;
- Helping to slow global warming and to mitigate climate change;
- Diminishing acid rain and promoting healthier forests, lakes, fish, and agricultural crops;
- Doing their part to reduce water pollution;
- Minimizing environmental devastation caused by mining, lumbering, and oil-drilling operations;
- Conserving increasingly scarce—and expensive—landfill space;
- Eliminating the need to build new power plants.

Re-using Is a Key to Creating a Sustainable Future—Because it conserves natural resources and cuts pollution, re-use offers an unsurpassed opportunity to people who want to tread lightly on Mother Earth—without lowering their standard of living.

Re-using means rethinking our role as "consumers" and becoming "conservers." It means stepping outside the prevailing culture from time to time and liberating ourselves from the thousands of daily messages on television, radio, magazines, and billboards exhorting us to spend and consume.

About This Book—The sections in this book are arranged in alphabetical order—from Appliances to Zippers. Each section describes ways that a specific product, such as coffee filters or personal computers, may be re-used. Most sections discuss how extending a product's useful life benefits the environment and saves you money. Many sections present success stories about individuals or businesses that have developed especially innovative or profitable approaches to re-use. And for readers who want additional information, most sections conclude with resource listings of helpful publications and organizations, as well as hard-to-find sources where re-usable products may be obtained.

Obviously, not all of the types of re-use described here are appropriate for everyone, though I hope many will strike you as so sensible or irresistibly economical that you'll want to give them a try.

The opportunities to re-use that I've described in this book represent only the tip of a great iceberg. An experimental spirit, a flash of Yankee ingenuity, and a spark of creativity can reveal dozens more ways to save money and resources by re-using.

Readers interested in further explorations may wish to consult the list of re-use oriented organizations in Appendix B, or the list of publications in Appendix C.

APPLIANCES

When I recently shelled out $18 to have the local fix-it guy install a new cord on my thirteen-year old toaster oven, my husband opined that I was probably wasting my money. "How long," he asked, "do you expect that thing to last?"

According to the folks at *Appliance* magazine, most people replace their toaster ovens after a mere three to nine years of service. I can't help wondering, however, just how many of those ousted ovens really died and how many were simply dumped for a sexy new model flaunting a newfangled convenience.

For $60, I could have acquired a shiny new toaster oven with sleek modern lines and a handy slide-out crumb tray. I could easily afford it, but could the Earth? Seeing how dearly our planet is paying for America's love affair with newness (acid rain, oil spills, open pit mines, and so on), I resolved to make my peace with the flecked and spattered patina of use on the appliance I already owned.

Was it worth it to have my toaster oven repaired? Only time will tell. I'm betting it will last for several more years. And if I'm right, the decision to repair will mean a little less damage to the planet and a little more money in my pocket.

Making Appliances Last Longer—The longer an appliance remains serviceable, the longer it stays out of the waste stream. It has been estimated that extending the service life of all household appliances by one-third would slow the discard rate by about 25 percent. That's 12 million fewer major appliances per year in the Great American Waste Stream.

In addition to saving significant tax dollars from avoided municipal waste disposal costs, longer-lived appliances would conserve the energy needed to manufacture replacements. Such conservation would diminish water pollution, urban smog, and acid rain. Less acid rain means healthier forests and crops. Less urban smog means healthier people. And less water pollution means, well, a healthier planet all the way around.

Maintenance Pays—Obviously, product design is a critical factor in determining appliance durability, but the way we care for and maintain our appliances also influences how long they'll serve.

Regular maintenance can maximize the service life of quite a few appliances, including refrigerators, furnaces, air conditioners, and even water heaters. Sometimes routine maintenance will enhance an appliance's energy efficiency. By cleaning the condenser coils on your refrigerator every few months, for instance, you can reduce electricity consumption 6 percent or more. (See ***Refrigerators;*** also ***Furnace Filters.***)

The average American home contains thousands of dollars worth of appliances. You can protect this investment by familiarizing yourself with the owner's manual and following the manufacturer's guidelines for care and maintenance. If you don't have an owner's manual, request one from the manufacturer or a dealer.

To Repair or Not to Repair—Timely repairs can often extend appliance life for many years. Unfortunately, the high cost of skilled labor frequently serves as a barrier to re-use when it discourages us from getting our ailing appliances fixed.

When the cost of repairs approaches an appreciable fraction of the cost of buying a replacement, many people will choose to discard the broken appliance and purchase a new one. In fact, one out of three readers who responded to the *Consumer Reports* 1992 annual questionnaire confessed that they hadn't bothered to fix major household appliances when they needed repairs; they simply bought new ones.

When one part wears out or an appliance stops working, people frequently assume that the appliance is on its last legs and that investing in repairs won't prove worthwhile. Appliance repairers and reconditioners assure me, however, that such despair may be premature and that plenty of discarded appliances are eminently repairable.

Appliance reconditioners often pick up the best of the trade-ins collected by dealers when selling replacement units. They clean up the discards, replace failed or worn parts, make necessary repairs, test, and resell them for roughly half the cost of comparable new ones.

Mr. or Ms. Fix-It—If you are an experienced tinkerer, you may be able to save a nice chunk of change by making repairs yourself. If you need assistance, ask the appliance manufacturer for help. Several major appliance manufacturers operate consumer help phone lines that can guide you through more complex repairs. You can request these numbers from a local dealer. (See ***Helpful Resources.***)

You can also obtain how-to manuals from local dealers or from a store that specializes in appliance parts. For simple repairs, some parts may come with all the instructions you'll need. General Electric, for example, sells "Quick Fix" kits for parts such as dryer belts, dryer vents, and ice-maker seals.

Face-Lifts—If a stove or oven functions fine but looks shabby, rather than replace it you may wish to have it professionally re-porcelainized by an appliance refinisher. As an alternative to refinishing, you may be able to purchase a new stove top (just the face panel without the burners) or a new oven door. Quite a few manufacturers also sell replacement panels and trim kits for recent model refrigerators and dishwashers. Ask a dealer or an appliance parts store about the availability of replacement parts.

It's also possible—and a whole lot less expensive—to spruce up the outside of a refrigerator with a special lead-free epoxy spray paint available at paint and hardware stores.

When Re-use Doesn't Pay—Sometimes it costs more to keep an old appliance chugging along than to buy a new more energy-efficient model. This is especially true with refrigerators and furnaces, because the new models guzzle so much less energy than older ones. Replacing a typical 18-cubic foot refrigerator manufactured in 1973 with an energy-efficient 1994 model, for example, can trim over 1,000 kilowatt-hours from your energy bill every year. Where I live, 1,000 kilowatt-hours costs approximately $120 (prices vary across the country). Over the fifteen-year service life of the typical refrigerator, that $120 annual savings will add up to $1,800—significantly more than the original purchase price.

Donate Instead of Dump—If your appliance functions well, but you decide to replace it anyway, direct the old one to someone who can use it. Charitable organizations such as Goodwill Industries welcome household appliances in good working order. Some nonprofit home rehabilitation programs like Habitat for Humanity or Christmas in April will also take working refrigerators, washers, and dryers. If you plan to declare the value of the appliance as a tax deduction, be sure to obtain a receipt from the recipient organization. (See Appendix A, ***Donate Instead of Dump.***)

Give It to a Reconditioner—If an appliance doesn't work, most charitable organizations have no use for it. Rather than drag it to the dump (or pay someone to haul it away for you), try offering it to a reconditioner, who may pick it up for free.

If you're looking to unload a dysfunctional appliance (or want to buy a reconditioned one), it's helpful to know that reconditioners usually specialize. Some deal exclusively in refrigerators, others stick to washers and dryers, or sewing machines, or vacuums. Look for them in the Yellow Pages under "Washers & Dryers," "Vacuums," and so on.

Purchasing New Appliances—When shopping for a new appliance, look for earth-friendly features such as low energy usage and a good repair history.

Energy-efficient appliances minimize environmental problems caused by burning fossil fuels—oil spills, acid rain, urban smog, and global warming, to name a few. Appliances with a reputation for infrequent repairs tend to suffer fewer unplanned breakdowns and can save you a bundle in avoided repair charges.

Federal law requires a bright yellow EnergyGuide label to -

appear on all new refrigerators, freezers, water heaters, dishwashers, clothes washers, room air conditioners, central air conditioners, heat pumps, furnaces, and boilers. These EnergyGuide labels tell you how energy efficient a particular appliance is when compared to other models of comparable size and type. The labels also show how much energy a particular model consumes in a year and translates that energy use into annual operating costs.

The appliance evaluations in *Consumer Reports* magazine often include a repair history for different manufacturers and models. These histories may help you select a durable appliance capable of staying out of the waste stream for a good long time. Look for *Consumer Reports* magazine in the reference section of your local public library. (Also see: ***Refrigerators*** and ***Washing Machines***.)

HELPFUL RESOURCES

1. Several major manufacturers have consumer-help phone lines staffed with technicians who can hold your hand through troubleshooting and/or repairs:

   General Electric, Hotpoint, Monogram, and **RCA** products: (800) 626-2000.
   Maytag products: (800) 688-9900.
   Whirlpool, Kitchen Aid, Roper, and **Estate** products: (800) 253-1301.

2. Updated annually, the ***Consumer Guide to Home Energy Savings*** lists the most energy efficient models of refrigerators, furnaces, water heaters, dishwashers, and clothes washers. I located a copy of this handy guide at the public library, but you may also purchase it in bookstores or directly from the publisher: **American Council for an Energy-Efficient Economy,** 2140 Shattuck Ave., #202, Berkeley, CA 94704; (510) 549-9914.

AUTOMOBILES

It's common knowledge that automobiles are bad news for human health and the natural environment. But we love the unlimited mobility our cars give us, so we've learned to tolerate the noise, pollution, physical danger, and expense that often come with them.

The good news is that a re-use strategy based on routine maintenance and thoughtful driving habits really can lessen a vehicle's destructive impact on the planet. Better yet, this same strategy, over time, can dramatically reduce the cost of owning and operating a car. If you can keep your car running smoothly and safely for a good long time, it is possible to chalk up some serious savings. As a car gets older, the fixed costs of owning it tend to go down; insurance premiums and registration fees are lower, and you don't have loan payments.

The secret of longevity for automobiles is regular maintenance. Proper maintenance also enhances fuel economy, helps ensure that your car is safe to drive, and can prevent some expensive repairs. Drivers who attentively maintain their vehicles can reduce both fuel costs and polluting emissions responsible for smog, acid rain, and global warming.

With the average new car priced at $20,000 in 1995 (and the average used car priced at more than $11,000!), it's more important than ever to keep your wheels in top condition for as long as possible. This would not be nearly so true, of course, if each year saw more fuel-efficient vehicles come to market. If new cars required significantly less gasoline than older models, the cost of buying a new vehicle might be offset by lowered operating costs. Until that happens, I'm going to make sure my car gets regular doses of preventive maintenance.

Maintenance—You don't have to be a mechanical genius to excel at auto maintenance. Just get your vehicle to a trustworthy mechanic and instruct him (or her) to perform the services prescribed in the owner's manual.

For a lasting and economical relationship with your car:

1. Change the oil (and filter) frequently. Dirty oil damages engine parts. Clean oil minimizes wear and thus extends engine life. Unless you drive under severe conditions, many mechanics recommend an oil change interval of 3,000 miles or 6 months, whichever occurs first. Check the owner's manual for your particular vehicle.
2. Check tire inflation at least once a month. Proper tire inflation saves gas (about 2 miles per gallon) and helps prevent excessive or uneven tire wear. For more detailed instructions on checking tire inflation, see ***Tires.***
3. Rotate tires every 6,000 miles.
4. Read the owner's manual for your car, and follow the manufacturer's recommended maintenance schedule. Every car comes with an owner's manual (it's often in the glove compartment), and you can always purchase one from the manufacturer or a local dealer. Most manufacturers sell manuals for vehicles up to ten years old. (See ***Helpful Resources.***)
5. At regular intervals (as prescribed in the owner's manual), check fluid levels such as automatic transmission, power steering, brakes, and coolant. Engines and transmissions can be damaged by insufficient fluids.
6. Flush the cooling system every two years (or as directed by the owner's manual). Flushing helps minimize accumulated rust and corrosion that might lead to an expensive radiator overhaul.
7. Change single-use air filters every 12,000 miles (more frequently if driving conditions are sandy or extremely dusty). If you have a washable/re-usable air filter, wash it at intervals recommended by its manufacturer.
8. Slow down. It has been estimated that fuel economy decreases as much as 25 percent when a vehicle is driven at 65 miles per hour instead of 55. Decreased fuel economy means increased pollution.
9. Drive less, walk more. This strategy saves wear and tear on

your car, cuts fuel costs, and tones your muscles. Who knows, it might even save you the cost of a gym membership.

HELPFUL RESOURCES

1. Unable to obtain an owner's manual for your car from a dealer? You can order one from **Helm, Inc.,** P.O. Box 07130, Detroit, MI 48207; (313) 883-1430.
2. Most libraries and book stores have plenty of books on car care. Two of my favorites are:

   Drive it Forever: Your Key to Long Automobile Life
   by Robert Sikorsky (McGraw-Hill, 1989, $9.95)
   The Planet Mechanic's Guide to Environmental Car Care
   by Jeff Shumway (B&B Publishing, 1993, $6.95)

3. Thinking about buying a used car? Each year in the April issue, ***Consumer Reports*** magazine publishes two very helpful listings of "Reliable Used Cars" and "Used Cars to Avoid." Most public libraries keep back issues of this magazine in the reference section.
4. The **U.S. Department of Transportation's Auto Safety Hotline** can tell you if a car model has ever been recalled and send specific information about the recall. (800) 424-9393.

BABY WIPES

For busy parents, disposable baby wipes are enticingly convenient, but they're also expensive and generate unnecessary waste.

As a simple low-waste alternative to baby wipes, dampened cotton cloths are unbeatable. Just toss soiled cloths in the diaper pail and launder along with diapers. When planning to be away from home with children, pack a damp washcloth in a clean plastic food bag (a re-used one, of course) turned right-side-out to avoid contact with printing that may contain toxic lead or cadmium. Re-usable snap-lock containers such as **Rubbermaid®** or **Tupperware®** work well too.

Over time, washcloths cost a whole lot less than throwaway wipes. And they save trees too. What's more, they don't contain drying alcohol or any of the other potentially irritating chemical ingredients listed on the commercial wipes containers.

Look for inexpensive cotton terry washcloths at baby specialty stores or order them by mail. (See ***Helpful Resources.***) Or make them yourself out of old towels.

If you must rely on disposable wipes, try purchasing them in refill packs. Drop-in refill packs cost less and use as much as 90 percent less packaging material than rigid plastic tubs. Drop-in refills fit neatly inside the heavier duty tubs, allowing the tubs to be re-used many times. Look for refill packs in drugstores and natural foods stores.

HELPFUL RESOURCES

Economically priced terry baby washing cloths may be ordered directly from:

Babyworks, 11725 N.W. West Rd., Portland, OR 97229; (800) 422-2910 or (503) 645-4349.

The Natural Baby Catalog, 816 Silvia St., 800 B-S, Trenton, NJ 08628-3299; (800) 388-2229.

BATTERIES

Hooked on Batteries—As a nation, we have a voracious appetite for batteries. To power our portable cassette players, radios, beepers, flashlights, calculators, and other gadgets, we purchase about 3 billion single-use batteries every year. Once they run down, virtually all of those batteries get tossed in the trash.

Since a single rechargeable battery can replace *several hundred* single-use alkalines, switching to rechargeables can dramatically reduce the number of batteries flowing into the waste stream. In addition to conserving pricey landfill space, using fewer alkalines reduces the consumption of energy and natural resources needed for battery manufacturing.

Rechargeables can save battery users a nice chunk of change too. According to researchers at Carnegie-Mellon University, operating a 4-battery cassette player for two hours a day over three years costs more than $650 if run on single-use alkalines. In contrast, powering the same cassette player with rechargeable nicads costs about $70 (including $55 for a state-of-the-art charger, $10 for four nicads, and $5 for the cost of electricity used for recharging).

Using Nickel-Cadmium Rechargeables—Although there are several types of rechargeable batteries, by far the most commonly used for household purposes is the nickel-cadmium cell. Known as "nicads," these batteries are real workhorses. With occasional time-outs for proper charging and a respectful appreciation of their idiosyncrasies, nicads can keep on going . . . and going . . . for up to 1,000 cycles.

1. Nicads work well in many, though not all, situations. The most common applications include portable tape players, flashlights, toys, radios, tape recorders, and beepers. I find they also work well in TV remote control devices.

2. Because the best nicads last only half as long between charges as single-use alkalines, it pays to shop around and compare amp ratings among major brands. The higher the amp rating, the longer the battery can go between charges. (See ***Helpful Resources.***)
3. To reach their maximum capacity, nicads must be properly "broken in." Experts recommend initiating nicads by fully charging and discharging them (letting them run down) five to eight times. Be sure to follow specific procedures described by the manufacturer.
4. If you play your personal stereo two hours each day while commuting to work, it's a good idea routinely to recharge the batteries at night—just as you replace an electric toothbrush in its charging unit after each use. If you buy two sets of cells, with one in the recharger and one in the tape player, you'll always have full power whenever you want it.
5. Nicads can be permanently damaged by overcharging. According to the battery experts at Real Goods Trading Corp., nicads can generally withstand being in the charger twice as long as necessary, but after that heat begins to damage them. Since charging times for nicads vary widely from three to twenty-four hours and sometimes longer, it is best to consult the manufacturer's literature for specific charging times.

   By far the easiest way to avoid frying your nicads is to purchase a "smart" charger that shuts off or cuts back to a tiny trickle when a battery is fully charged. (See ***Helpful Resources.***)
6. Always use the type of charger specified by the battery's manufacturer.
7. Because they lose 1 percent of their charge every day (even when you don't use them), nicads are not suited for use with clocks or watches or for standby applications such as smoke alarms, garage door openers, or cameras.
8. Electronic devices that demand exactly 1.5 volts of power should not be used with nicads, which deliver only 1.25 volts. Be sure to check the product manufacturer's instructions.

Using Alkaline Rechargeables—For devices that need 1.5 volts of power, the **Renewal®**, a special rechargeable alkaline battery that delivers the requisite 1.5 volts, may be just what the doctor ordered.

According to the manufacturer, Renewals can be recharged 25 times and last up to three times longer per charge than nicads. Unlike nicads, which lose 30 percent of their power per month, Renewals retain their power in storage for up to 5 years.

Even though Renewals cannot sustain anywhere near as many recharges as nicads, they are less expensive and more resource efficient than 25 rounds of single-use alkalines.

Disposing of Rechargeable Batteries—Renewals are mercury free, so householders can toss them in the household trash with a relatively clear conscience. (Under federal law, Renewals are not considered "hazardous household waste.") It would be best, of course, to recycle these batteries, but economically viable recycling technologies for alkalines are not expected to come on line before the turn of the century.

Spent nicads should not be tossed in the household trash. Because they contain about 20 percent cadmium (a toxic heavy metal known to cause lung and kidney diseases and possibly cancer), nicads should be disposed of as hazardous household waste or, preferably, recycled. (See ***Helpful Resources.***)

The Rechargeable Battery Recycling Corporation (RBRC) is in the process of establishing a national collection program to recycle spent nicads. Financed with licensing fees paid by the manufacturers of nicad batteries and battery-powered tools containing built-in nicads, RBRC provides free collection containers to community hazardous waste programs and retail stores that sell nicads. When full, the containers are forwarded to a state-of-the-art recycling facility in Elwood City, Pennsylvania. As of this writing, the collection program for nicads is up and running in seven states. By 1998, the program is expected to be operational in all fifty states. (See ***Helpful Resources.***)

HELPFUL RESOURCES

1. Which nicad brands are best? ***Consumer Reports*** magazine tested several major brands and published the results in the December 1994 issue. Look for back issues of *Consumer Reports* in your public library's reference section.
2. ***The Real Goods Catalog*** guarantees its high-capacity nicads for the purchaser's entire lifetime. **Real Goods** product information specialists can answer all your battery questions: 555 Leslie St., Ukiah, CA 95482-5507; (800) 762-7325.
3. "Smart" chargers may be purchased from the **Real Goods Trading Corp.,** 555 Leslie St., Ukiah, CA 95482-5507; (800) 762-7325.
4. If you'd like more information about **Renewal® Reusable Alkaline Batteries,** contact the manufacturer, **Rayovac Corporation** at P.O. Box 44960, Madison, WI 53744-4960; (800) 237-7000.
5. Want to recycle dead nicads? For the nearest recycling collection site, call the **Rechargeable Battery Recycling Corporation's** toll-free information helpline: (800) 8-BATTERY.
6. To find out if your community has a collection program for hazardous household wastes, call a local recycling agency, the city sanitation department, or the health department. If your city or county does not have such a program, the EPA publishes a free manual that explains how to set one up. To obtain a copy of ***Hazardous Household Waste: A Manual for One-Day Community Collection Programs,*** call the EPA's information hotline at (800) 424-9346.

BEVERAGE CONTAINERS

Where Have All the Refillable Bottles Gone?—In many European and South American countries as well as in neighboring Canada, the practice of refilling beverage bottles is alive and well. But in much of the United States, refillable bottles have essentially become a thing of the past.

For me, the sight of a well worn Coca-Cola bottle made of pale green glass provokes a fit of nostalgia, recalling vacation trips to the mountains with pit stops at gas stations where the vending machines dispensed sodas in thick glass bottles, and you were expected to replace the bottle on a rack of empties before driving on.

In 1950, most of the beer and soda bottles in this country were refilled 10 to 30 times. Coke bottles averaged 50 refillings. No wonder they had that splendidly abraded patina of use. Since the 1960s, however, refilling has steadily declined as single-use aluminum cans and lighter weight glass bottles have come to dominate the beverage container market. By 1991, less than 9 percent of America's beer and soft drinks were sold in refillable bottles.

Bucking the Trend—A growing number of local bottlers are discovering that the logistics of returning empties for refilling can be enticingly economical. The New England Brewery in Norwich, Connecticut, for example, reported that retrieving and refilling used bottles costs about half as much as buying new ones.

The Schroeder Milk Company in St. Paul, Minnesota, passes the savings on to its customers, selling milk in refillable plastic jugs for 8 to 16 cents less per gallon than milk packaged in throw-away containers. And in eastern New York state, Stewart's convenience stores sell private label sodas in refillable glass bottles for half the price of national brands.

In Stockton, California, the Ever Green Glass company uses an automated processing line and electronic scanner to wash and sterilize discarded wine bottles. These bottles are much in demand among the region's wine makers, because they cost 10 to 15 percent less than new ones.

Refillables Save Energy—It's a fact: Refillable glass bottles are the most energy efficient single-serving container currently available. If glass bottles are refilled an average of 10 times, about 75 percent less energy is required per use than to manufacture aluminum cans or glass bottles from recycled ones.

Twelve months after switching to refillables, the Rainier Brewing Company of Seattle and Blitz-Weinhard in Portland, Oregon, reported having saved enough energy to serve 3,441 homes for a year.

Refillables Strengthen Local Economies—According to Scott Chaplin, a senior researcher at the Rocky Mountain Institute, refillables actually create jobs, because bottling plants that wash and refill require more labor than those that use one-way containers.

Refillables Reduce Solid Waste—A growing number of dairies now deliver milk to schools in refillable bottles made from polycarbonate, a plastic resin that is clear like glass but much less breakable and which can be washed and refilled up to 100 times. GE Plastics, which manufactures refillable milk jugs and bottles, estimates that if all the schools in New York City switched from paper cartons to refillable plastic milk containers, New York City could save at least $500,000 per year in trash removal costs.

The Future of Refillables—In spite of an encouraging number of local success stories, the United States has a long way to go before matching the levels of refilling achieved in other countries.

According to the Brewers Association of Canada, over 97 percent of the beer bottles used in Canada are returned for refilling. In Holland, 90 percent of all beer bottles are refilled. In

Mexico, refillables account for 80 percent of the beer container market and more than 70 percent of soft drink container market. And in Japan, 97 percent of beer bottles and 83 percent of sake bottles are refilled.

Do We Need a New Law?—It is often legislation, combined with hefty bottle deposits, that provides the necessary impetus for successful national refilling systems. In Denmark, where officials estimate that 99 percent of beer bottles are returned for refilling, one-way soft drink and beer containers have been banned since 1981. And in Norway, where refilling rates approach 90 percent, the government levies a special product charge on all single-use containers.

Vote with Your Pocketbook—The success of fledgling efforts to bring back refillables will depend upon supportive patronage by ecologically aware consumers. If your local grocery store doesn't offer beverages in refillables, ask the store manager to stock them. When beverages *are* available in refillable containers, make a point of purchasing them. And be sure to return the containers!

Make a Free Phone Call—Check the label on your favorite beer or soft drink for a toll-free consumer information number. Phone the manufacturer and request beverages in refillable bottles.

Coca-Cola, Hi-C, Minute Maid, Sprite, Tab, Mr. PiBB, Fresca, Nestea iced teas, **POWERaDE, Fruitopia, Barq's** root beer, **Fanta** 1-800-438-2653
Pepsi-Cola, Dr. Pepper, Schweppes, Slice, Mug Root Beer, Lipton's Brew, Seven-Up, Canada Dry, Mountain Dew, Mirinda Orange 1-800-433-2652
Shasta brand soft drinks 1-800-643-0745
Hawaiian Punch 1-800-477-8624
Ocean Spray 1-800 662-3263
Snapple 1-800-SNAPPLE

Crystal Geyser, Juice Squeeze 1-800-4-GEYSER
AriZona Tea beverages 1-800-TEA-3775
Miller Beer 1-800-MILLER-6
Coors Beer 1-800-642-6116
Anheuser Busch (Budweiser, Michelob) 1-800-342-5283

Write a Short Letter—Urge state and national lawmakers to sponsor legislation that will encourage the use of refillable beverage containers.

Make Your Own Soft Drinks—If it's caffeine you crave, how about making your own iced tea or coffee? Homemade beverages cost much less than store-bought ones. If you must have carbonation, you can create your own sodas with a **Spritzit Siphon System,** a handy gadget that lets you prepare your own favorite sodas (Coke, Pepsi, Seven-Up, etc.), seltzer water, and sparkling juice drinks for about 15 cents per liter. (See ***Helpful Resources.***)

HELPFUL RESOURCES

The **Spritzit Siphon System,** which lets you make low-cost carbonated drinks at home, utilizes refillable carbon dioxide cylinders. For a free brochure, contact **Globus Mercatus, Inc.,** at 12 Central Ave., Cranford, NJ 07016; 1-800-NATURE-1.

BICYCLES

The Rewards of Regular Maintenance—The key to a long and satisfying relationship with your bicycle is, you guessed it, regular maintenance. Periodic cleaning and lubrication can make your bike a joy to ride, allowing you to get the most speed for the least effort. A bike that is attentively cared for is also safer—and cheaper—to operate because frequent inspections and adjustments often nip accident-causing mechanical problems in the bud and prevent expensive repairs.

Most preventive maintenance consists of simple activities such as cleaning, lubricating, checking for loose parts and tightening them up as needed. A good maintenance and repair manual can introduce you to the basic procedures and guide you in selecting the tools, lubricants, and cleaning materials you'll need. (See ***Helpful Resources***.) If you prefer hands-on learning supervised by an expert, contact local bicycle shops or bike clubs and ask about classes in maintenance and repair.

Recycled Bicycles—Got an old bike you don't want or that is badly in need of repair? More than one hundred bike re-use programs across the nation will be happy to take it off your hands. The Re-Cyclery, a bicycle thrift shop in San Rafael, California, is an outstanding example of what such a program can accomplish.

At the Re-Cyclery, which is much more than a thrift shop, bicycle mechanics teach free after-school bike repair classes to low-income kids aged ten to seventeen. For each time they show up and participate, the kids earn $1 credit toward the purchase of a used bike.

The Re-Cyclery is the brainchild of Marilyn Price, who also organizes an annual bike swap that draws hundreds of cyclists from around the San Francisco Bay Area. Proceeds from the bike swap and Re-Cyclery help to finance Trips for Kids, a volunteer program that takes inner city youths on weekly mountain bike excursions.

In addition to old bicycles donated by individuals, Ms. Price receives stolen bikes confiscated by local police departments and the county sheriff. Vehicles in good working condition are sold at the annual bike swap or at the Re-Cyclery. Those in need of repair are delivered to nearby San Quentin prison, where inmates "cannibalize" dysfunctional bikes to create working bikes. Little goes to waste or ends up in landfill. The refurbished bikes are donated to charities, and the scrap metal from unusable parts is recycled.

HELPFUL RESOURCES

1. For cyclists who would like to learn basic maintenance procedures, ***Bicycle Magazine's Complete Guide to Bicycle Maintenance and Repair*** (Rodale Press, 1994, $16.95) provides step-by-step instructions and plenty of illustrations.
2. The **Re-Cyclery** and **Trips for Kids** are admirable programs, but they are hardly unique. Similar programs are popping up across the country. The **Youth Bicycle Education Network** (YBEN) promotes the spread of similar programs to communities across the country. Individuals interested in starting such a program can consult experts who already operate successful bike education programs and obtain start-up manuals through YBEN, which also publishes an informative quarterly newsletter. Contact the Youth Bicycle Education Network, YBEN/SCCCC, P.O. Box 8342, Santa Cruz, CA 95061- 8342; (408) 457-2027.

BOOKS

Most people use the public library for economic reasons. (Books are expensive, and the library is free.) But there are also plenty of sound environmental reasons to borrow books rather than buy them.

By systematically enabling an entire community to share a single book, America's public libraries eliminate the need for millions of individual copies. This may not be good news for publishers and royalty-dependent authors, but it is excellent news for the earth and its inhabitants.

Without ever intending to, libraries over the years have conserved vast tracts of forest. They have also saved prodigious amounts of energy, preventing a good deal of noxious air pollution and combating global warming. By curtailing the demand for paper products, libraries have unwittingly abated highly toxic chemical discharges from the paper industry's pulp plants.

Of course, libraries aren't the only book-sharing system in town. Secondhand book stores, library fund-raising sales, garage sales, antique stores, and thrift shops facilitate millions of book swaps every year. Some recycling centers even have a book shelf where recyclers can leave unwanted volumes and browsers can take home free books.

Whenever I run out of shelf space and need to shed a few books to make room for more recent acquisitions, I sell my discards to used book stores or donate them to the local library's book sale committee. For the many public libraries that hold annual book sales to raise funds for library materials, donated books are a precious resource. When you donate, don't forget to request a receipt so you may claim a charitable deduction for the fair market value of your gift. (See Appendix A, ***Donate Instead of Dump.***)

School librarians often welcome used books in good condition, provided they are age-appropriate. It's best to contact the library first to ask what sort of books they want. Senior centers frequently accept all types of literature, current as well as classic. Large print books are especially in demand.

Building Community Libraries Overseas—Obviously, the most energy-efficient way to share books is to exchange them locally. This saves shipping costs and conserves the energy required for long-distance transport. Nevertheless, there is a burgeoning demand in developing countries around the globe for books and periodicals. To meet these needs, several nonprofit organizations have established programs for sending books abroad. (See ***Helpful Resources.***)

HELPFUL RESOURCES

The **International Book Project** is dedicated to building community libraries overseas. This group needs textbooks, technical books, and reference books in good condition. In general, books should be less than ten years old and have bindings sturdy enough to endure five years of hard use. The Project's needs are very specific, so please contact them first for book selecting instructions. International Book Project, 1440 Delaware Ave., Lexington, KY 40505; (606) 254-6771.

BUILDING MATERIALS

It's happening. Though long overshadowed by recycling, re-use is gaining ground. Every day, more and more Americans are utilizing discarded materials in creative and resourceful ways. If you don't believe me, just find a used building materials store (UBMS) and pay it a visit.

During the last five years, UBMSs have been springing up across the country like mushrooms after a spring rain. According to *The Re-User,* a newsletter for the recently formed Used Building Materials Association, the number of UBMSs in the U.S. and Canada more than doubled between 1989 and 1994. If this trend continues, within ten years UBMSs will be established in hundreds, possibly thousands, of North American communities. A 1994 survey by EarthWorks Environment indicates that already there are nearly 200 UBMSs in Canada and the United States.

Mining Urban Wastes—Business was brisk on the day I visited Urban Ore, a used building materials store in Berkeley, California. A steady stream of customers from all walks of life browsed among the neatly ordered rows of forsaken bath tubs, sinks and toilets, vintage and modern doors, windows, kitchen cabinets, miscellaneous lumber, hardware, and bricks.

Urban Ore purchases these materials from contractors and individuals, then sells them "as is" to the general public. Among its customers, Urban Ore counts remodel contractors and interior designers, collectors and antique dealers, landlords and homeowners, building maintenance people and home repairers, artists and inventors, housewives, teachers, and gardeners.

According to company president and founder, Dan Knapp, more than 95 percent of the incoming goods are sold for re-use. Materials that cannot be re-used are often scrapped for recycling. Less than one-half of one percent goes to landfill.

The City of Berkeley has estimated that in 1990 Urban Ore

diverted more than 5,000 tons of materials from the local waste stream—as much material as the city's curbside recycling program collected that year in cans, bottles, and newspapers.

Boosting the Local Economy—Urban Ore exemplifies the ways in which redistributing re-usable discards can animate a local economy. In the process of grossing close to $1.5 million in 1995, the company created jobs, collected sales tax for state and local jurisdictions, paid local people who sold their goods, and sold salvageable goods to local crafts and tradespeople who earned income from the value they added through repairing and refurbishing. In turn, these reconditioners provided area residents with usable goods more affordable than "brand new" ones.

The Science of Salvaging—After more than fifteen years of operation, the folks at Urban Ore have the business of mining urban wastes down to a science. Eager to share with others what they have learned, the company distributes technical papers and a slide show on how to manage a profitable used building materials store. (See ***Helpful Resources***.)

Relishing the Good News—The environmental impact of channeling discarded building parts to new users is profoundly positive. Re-using any sort of building material—from kitchen cupboards to fireplace bricks—conserves the natural resources, energy, and human skills embodied in those items. Re-using old lumber, for instance, saves more than just trees; it conserves the energy needed to chop them down, process the logs into products such as buildable boards and finished doors, and transport these products to urban building sites.

How You Can Get Involved—Anyone who has ever tackled a home remodeling project knows that dump fees can easily add up to several hundred dollars. Taking re-usable items to a UBMS is a great way to trim disposal costs. (To help ensure that your discards can be re-used, try to dismantle the existing construction as gently as possible.)

If the UBMS that accepts your discards is a for-profit business, you may receive a modest payment for your stuff. If it's nonprofit, you could reap a nice tax deduction. Either way, you'll save money on avoided dump fees and extend the life of local landfills.

Look for UBMSs in the Yellow Pages under headings such as "Building Materials—Used"; "Lumber—Recycled"; "Building Materials—Architectural, Antique, and Used"; or "Salvage."

If your area doesn't yet have a UBMS, you may wish to explore these options:

1. Offer re-usable items in good condition to a home rehabilitation program such as Habitat for Humanity. Such programs sometimes accept tools, sinks, cabinets, gas stoves, and light fixtures in good condition. Please call first to confirm need. If you can't find a Habitat affiliate in your area, ask a local church to direct you to a home rehabilitation project.
2. Offer unwanted building materials to your neighbors. One couple whose California home was seriously damaged by the 1989 Loma Prieta earthquake invited their neighbors to salvage whatever they wanted before the demolition contractor arrived. Dozens of grateful neighbors hauled off everything from light fixtures to banister railings. Even the stove and dishwasher found happy new owners.

(Also see: ***Lumber, Houses.***)

HELPFUL RESOURCES

1. The **Used Building Materials Association** can direct you to a UBMS in your area: #2-70 Albert St., Winnipeg, Manitoba, Canada R3B 1E7; (204) 947-0848.
2. For a free list of California businesses that buy and sell used building parts, contact the **Waste Prevention Info Exchange,** 8800 Cal Center Dr., Sacramento, CA 95826; (916) 255-INFO.

3. **Urban Ore Information Services** answers inquiries and distributes the company's publications, which include technical papers and a slide show on how to set up and operate a profitable used building materials store. Contact Urban Ore Information Services at 1333 6th St., Berkeley, CA 94710; (510) 559-4454.
4. In pursuing its mission to build simple decent housing for low-income families, Habitat for Humanity often needs home furnishings and building materials in good condition. Check the phone book for a local office or contact **Habitat International,** 121 Habitat St., Americus, GA 31709; (800) HABITAT or (912) 924-6935.
5. Since the 1970's, builder-architect Michael Reynolds has built or designed over 500 homes using discarded tires and aluminum cans. Books and videos describing Reynolds's home-building techniques may be obtained from **Solar Survival Architecture,** P.O. Box 1041, Taos, NM 87571; (505) 758-9870.
6. An excellent book for anyone intending to utilize used building materials is ***Building with Junk and Other Good Stuff: A Guide to Home Building and Remodeling Using Recycled Materials*** by Jim Broadstreet (Loompanics Unlimited, 1990, $19.95). Available in bookstores or directly from the publisher, **Loompanics Unlimited,** P.O. Box 1197, Port Townsend, WA 98368.

CARPETS

Though there are no official surveys or figures, some carpet industry professionals estimate that as much as 20 percent of all discarded carpeting eventually finds a second home. One explanation for the lack of definitive numbers is the fact that many of the outlets which channel old carpets to new owners are part of a low-profile "grey" market that doesn't invest much in advertising and therefore may require a bit of consumer sleuthing to discover.

Searching for Used Carpet—"OK, suppose I want to buy a good-sized piece of used carpet for my basement. Where would I find such a thing?" That is the question I put to Pete Hovde, the founder of Carpet Renovation, a company in Princeton, Minnesota that contracts with property owners and others to remove old carpets from office buildings, hotels, and resorts.

"Carpet Pete," as his customers call him, estimates that he finds new owners for about 70 percent of the carpets he removes. He sells the carpet "as is" for $1 to $2 per square yard—approximately one-tenth what it could cost to buy new. The 30 percent that doesn't find a new home usually ends up in a Minnesota landfill, though Pete is pretty sure he could probably sell about half of those now being landfilled by cleaning and/or dying them. He hopes to test his hunch soon; last year, he took a course in professional carpet cleaning and dying.

Although a handful of the largest used carpet dealers advertise in the Yellow Pages, Carpet Pete doesn't. It's too expensive. Pete has done some local advertising in the past, but no longer finds it necessary. "The word is out," he told me. "Some people come from 100 miles away to buy used carpet from me."

Pete suggested I try contacting carpet installers in my area, since they are the ones who usually get hired to remove the old carpet. "I'm not the only one in the country who is doing this," he insisted. "I know installers who are selling used carpet in Phoenix and Denver, so I'm sure it's being done all over the country."

The next time I found myself perusing classified ads in the local advertiser, a tiny two-line ad for "carpet installation and cleaning" caught my eye. I phoned the installer who ran the ad, and he told me he had several large pieces of used carpet in excellent condition. He had cleaned them himself, restretched them, and repaired the flaws. "They would be perfect," he assured me, "for your basement."

Carpet installers are not the sole source of secondhand carpeting. Used building materials stores are also good places to seek out such stuff. (A newcomer to the urban scene, used building materials stores are discussed at length under ***Building Materials.***)

Are Used Carpets Less Toxic Than New Ones?—One often overlooked advantage of secondhand carpeting is that it may be relatively free of the dreaded "new carpet smell" which has been linked to a variety of disturbing symptoms ranging from lightheadedness and difficulty in concentrating to numbness, nausea, dizziness, and double vision.

Researchers now believe that the chemical smell so characteristic of much new carpeting is probably caused by the offgassing of volatile organic compounds (VOCs). Although scientists do not yet fully understand why some new carpets make some people sick, researchers suspect that latex backing may be the culprit.

Until the offending chemicals are definitively identified and carpet manufacturers take steps to eliminate them, used carpeting may provide a less toxic alternative, since it may already have emitted many of its VOCs in its previous installation. According to Debra Lynn Dadd, author of *Nontoxic, Natural & Earthwise,* if a carpet is more than five years old it is likely that most of the toxic VOCs have already gassed out.

Making It Last—If you want to get your money's worth out of carpeting, there are steps you can take to help ensure it lasts a good long time.

1. Regular vacuuming (at least once a week) is essential. Vacuuming can minimize the danger that ground-in dirt particles

will abrade carpet fibers, chewing into the fabric of the carpet as it is walked on.
2. Carpet manufacturers recommend deep cleaning carpets every twelve to eighteen months.
3. Wipe up spills promptly. The longer a spill remains on the carpet the more difficult it is to remove and the more likely it is to leave a stain. Even carpet that has been treated with a stain resister is not "stain proof."
4. To prevent premature fading, draw drapes to block intense afternoon sun.

Getting It Fixed—Just because it incurs a spot of damage doesn't mean it's curtains for an entire roomful of carpet. Professional carpet repairers and dyers have a repertoire of restorative techniques that include: redying to cover stains and fading, repairs to cigarette burns, rips and tears, pet accidents, split seams, and water damage.

Do-it-yourselfers who want to make carpet repairs will need the right tools, and these can be purchased in a carpet store or from a carpet layers equipment and supply store. Look for books containing carpet repair instructions at the library or bookstore. (See ***Helpful Resources.***)

HELPFUL RESOURCES

1. To obtain a free ***Carpet Owner's Manual*** containing long-term care instructions and detailed advice on spot removal, contact the **Carpet and Rug Institute,** P.O. Box 2048, Dalton, GA 30722-2048; (800) 882-8846.
2. ***Floors, Stairs & Carpets*** (Sunset Publications, 1994, $12.99). Provides step-by-step instructions for diagnosing and repairing basic carpet problems.
3. Want to know more about new carpet smell and how to avoid it? Contact ***The Green Guide,*** published by **Mothers & Others for a Livable Planet,** 40 West 20 St., New York, NY 10011-4211. Request Publication #19 (January 7, 1996), "Carpet: Laying it Safe."

CHILDREN'S CLOTHING

In 1980, when her daughter was six months old, Karen Lynch had a pile of already outgrown clothes and equipment worth about $2,000. That's when she came up with an idea whose time appears to have come: A clean well-lighted resale boutique where budget-stressed parents can purchase high quality like-new children's clothing, toys, and equipment for a fraction of their original cost; and where parents can unload lightly-used items for cash.

Ms. Lynch's store, which she named **Children's Orchard®**, was so successful that in 1985 she began franchising the idea to aspiring entrepreneurs in other cities. Over the last ten years, she has sold 55 franchises in eleven states.

It's a Trend—In addition to franchises, plenty of independent resale boutiques have been popping up all over the country, and many are thriving. It makes sense. Healthy young children outgrow costly clothing every few months or so, and today's smaller families provide few hand-me-down opportunities.

Bringing Down the Cost of Growing Up—Some resale shops pay cash for used items; some pay in credit toward the purchase of lightly used merchandise; others work on a consignment basis and pay a percentage of the sale price only when goods are sold.

Most of these resale boutiques handle only gently used gear of the highest quality. If the upscale outlets reject your child's togs and toys because they show greater signs of wear, plenty of charity thrift shops will be delighted to receive them. And the savvy moms who bargain hunt at thrift shops will be glad to buy them. If you want to claim a tax deduction for your donation, be sure to request a receipt. (See Appendix A, ***Donate Instead of Dump.***)

Safety First—To avoid purchasing children's sleepwear that might have been treated with TRIS (a flame retardant banned since the late 1970s and believed to cause skin cancer), the Consumer Product Safety Commission advises parents not to purchase items made from acetate, triacetate, or blends of these two materials. The Commission also recommends avoiding any children's sleepwear with labels removed. Always read garment labels carefully.

A Stitch in Time—A friend of mine once lamented that by the time she got around to mending her kids' clothes, they had often outgrown them. Too bad. She missed some great opportunities to save resources, financial as well as natural. Many a simple mend has extended the useful life of a child's garment and spared parents the expense of buying a new outfit that would soon be outgrown anyway.

HELPFUL RESOURCES

1. ***Singer® Children's Clothes, Toys & Gifts*** (Cy DeCosse, 1995, $15.95) provides clear and beautifully illustrated instructions on repairing children's clothes and sewing simple, durable clothes for infants and toddlers. Available in bookstores or directly from the publisher, **Cy DeCosse, Inc.,** 5900 Green Oak Dr., Minnetonka, MN 55343; (800) 328-3895.
2. These companies franchise children's resale boutiques:

   Children's Orchard, 315 E. Eisenhower, Suite 316, Ann Arbor, MI 48108; (800) 999-KIDS or (313) 994-9199.

   Once Upon a Child® (Grow Biz International), 4200 Dahlberg Drive., Minneapolis, MN 55422; (800) 445-1006 or (612) 520-8500.

CHILD SAFETY SEATS

Anyone who frequents garage sales knows that there are plenty of previously owned car seats for infants and small children out there, and the prices are tantalizingly low. But don't let price be your sole consideration; safety is very important. Here are some tips to help you find a seat that delivers the best protection.

1. Avoid car seats manufactured before 1981, when the federal government set more stringent safety standards for child safety seats. Make sure the stamp of manufacturing is dated after January 1, 1981. Look for this label on the safety seat: "This child restraint system conforms to all applicable Federal motor vehicle safety standards."
2. Before you buy, check to see whether the car seat has been recalled. According to *Consumer Reports* magazine, child safety seats are among the most commonly recalled consumer products. You can easily check the recall status of a specific model by calling the National Highway Traffic Safety Administration's toll-free Auto Safety Hotline. (See ***Helpful Resources.***) Before calling, you'll need to know the brand, model, and date of manufacture. This information appears on every seat.

   Upon request, the Auto Safety Hotline will send a current list of child seat safety recalls, handy to have along when you're making the rounds of garage sales or resale boutiques.
3. Most recalls don't require replacing the entire seat, and manufacturers will often send free substitute parts with replacement instructions when consumers request them. If the recall action specifies "Replacement," however, you could receive a brand new seat from the manufacturer.
4. After purchasing a used child safety seat, contact the manufacturer and request a registration card. That way, you'll be notified in the event of a recall.
5. Always follow the manufacturer's instructions for installation and use. If instructions are missing, contact the manufacturer

and request a copy. Only a properly installed seat can deliver the protection it was designed to provide.

6. Both cars and children come in different sizes. Make sure that the car seat is appropriate for your child as well as for your car. If it doesn't fit, don't buy it.

Psst! Pass It Around—If the car seat is still sound when your child outgrows it, don't trash it. Provided it meets the safety criteria outlined above, pass it on to a friend or donate it to a charitable organization like Goodwill Industries or Salvation Army. (See Appendix A, ***Donate Instead of Dump.***)

HELPFUL RESOURCES

The American Academy of Pediatrics has prepared a ***Shopping Guide for Child Safety Seats.*** This guide is available for free, along with a list of current car seat recalls, from the **National Highway Traffic Safety Administration, Auto Safety Hotline:** (800) 424-9393. Or for Washington, D.C. area callers only: (202) 366-0123.

If you prefer, you may send a written request for a "Child Safety Seat Packet" to: **National Highway Traffic Safety Administration, Auto Safety Hotline,** NEF-112 HL, 400 Seventh St., S.W., Washington, D.C. 20590.

CHRISTMAS TREES

You Have Choices—If you really want a conventional cut tree for Christmas, for heaven's sake go ahead and get one. There's no need to wallow in guilt about it, because most modern Christmas trees are grown on tree farms and are therefore considered a "renewable" resource. Of course, when the holidays are over, you'll want to dispose of your tree in a way that doesn't burden the environment—by recycling it at a municipal composting facility or by composting it yourself.

On the other hand, if you're distressed about the fact that the world's forests are falling at the alarming rate of fifty acres per minute and you just don't feel right about celebrating the holiday of joy and hope by chopping down yet another tree, there are plenty of festive alternatives that don't involve dead trees.

"Green" Variations on the Traditional Tree—John Muir's favorite way to decorate the house for Christmas was to drape garlands of evergreen cuttings over the door lintels. At my house, we "deck the halls" by arranging cedar branches pruned from a tree in the front yard on the fireplace mantel. I nestle a couple of tall red candles and some tiny ornaments in among the greenery, then savor the scent of freshly cut cedar.

With the addition of a few simple ornaments, a large indoor plant or tree can become a yuletide work of art. Last year, a neighbor of mine adorned the giant ficus tree in her dining room with tiny blinking lights. It was an enchanting sight. If you decide to try this, use only small low-wattage lights like the new "cool" 5-watt lights. And avoid lights altogether on sensitive plants that may be damaged by the heat that lights give off.

Living Evergreen Trees—A potted evergreen tree can be brought indoors at Christmas time and redecorated season after season. Several years ago, my sister invested in a potted dwarf Alberta Spruce, which she plans to use for the next seven holiday seasons.

She figures that this $14 four-footer will save her about $336 over the cost of a somewhat taller "dead" tree for which she'd have to pay $50 every year.

Living trees generally shed fewer needles than dead ones. However, since the air in heated homes tends to be dry and inhospitable to live trees, they should remain indoors for not more than ten days. A living tree will do best in a cool part of the room, away from heat sources. Water it sparingly every three days—just enough to keep the soil moist. Provided your tree isn't decorated with lights, it's OK to mist it occasionally, but *never* spray water on a plant with lights. Small low-wattage lights are OK; most other lights will burn the needles.

According to the Evergreen Alliance, the twelve most popular species for living Christmas trees are Douglas Fir, Noble Fir, White Fir, Norway Spruce, Colorado Blue Spruce, Scotch Pine, Norfolk Island Pine, Japanese Black Pine, Aleppo Pine, Monterey Pine, and Mondell Pine. Ask a reliable nursery to recommend one that will thrive in your climate.

If you intend to re-use your living tree for many seasons, select one that grows slowly. A potted White Fir, for instance, will grow only 1 to 4 inches yearly, while a Douglas Fir will shoot up 1 or 2 feet. Dwarf Alberta Spruce trees *never* grow tall, adding only 4 inches a year when they are young and even less in later years.

To maintain your living tree between yuletides, put it in a hospitably sunny outdoor spot. (If it freezes outside where you live, ask a nursery how to protect the root ball.) Water the tree regularly. Fertilize it two or three times during spring and summer.

When the roots of your tree become cramped (every two to three years), you'll need to repot it in a slightly larger container. Transplant the tree in firmly packed soil at the same depth at which it stood in the previous container. A reliable nursery can tell you what type of potting soil to use and the best time of year to transplant. If you want to slow the tree's growth, however, you won't need to buy a new pot at all. You can remove the tree from its pot, prune the roots, and then repot it in the same container with some fresh soil. Root pruning is a simple procedure. Ask a nursery for instructions or consult a book on container gardening. (See ***Helpful Resources.***)

"Fake Firs"—Because it was "too much work and too messy," my mother-in-law gave up her cut trees over a decade ago. She opted instead for a re-usable artificial tree. After the holidays, she stashes the whole caboodle—lights, ornaments and all—in her basement and covers it with an old sheet until next December.

According to the National Christmas Tree Association, about half the trees displayed each year in the U.S. are artificial firs. Although these trees don't have the woodsy scent of real pine, fir, or spruce, they won't drop needles all over the living room floor either, and they pose less of a fire hazard than dead, cut trees. The Asthma and Allergy Foundation recommends artificial trees to allergy sufferers, because real trees are often laden with pollens and molds that can cause sneezing, wheezing, congestion, itchiness, and watery eyes.

Safety First—Whatever kind of tree you have, never leave the house or go to sleep with the tree lights on! Smaller bulbs may be safer because they give off less heat. They also consume less electricity and cost less to operate.

HELPFUL RESOURCES

1. Prepared by **The Evergreen Alliance, *The New Green Christmas: How to Make This and Every Holiday an Environmental Celebration*** (Halo, 1992, $5.95) provides a wealth of advice on the care and selection of re-usable evergreen Christmas trees.
2. ***Container Gardening*** (Sunset, 1984, $9.99) provides easy instructions on root pruning.

CLOTHING

Clothing re-use has actually become quite popular lately, if not downright chic. America's bargain hunters are snapping up secondhand clothes at an expanding network of resale stores, thrift shops, yard sales, and upscale consignment boutiques. Meanwhile, textile recyclers report that supplies of used clothing are increasingly scarce and observe that Americans are buying fewer clothes and wearing them longer.

One of the beauties of clothing re-use is that virtually everyone can participate. Even mad shoppers and "clotheshorses" can be part of the re-use process when they channel their surplus garments to resale outlets.

Tax Deductions for Donations—If you choose to donate to a charitable organization, you may receive a tax deduction for the fair market value of your donation. Organizations that appreciate receiving your old clothes include charity thrift shops, social service organizations, homeless shelters, and women's shelters. Some churches also welcome used clothing donations. (See Appendix A, ***Donate Instead of Dump.***)

Don't Be a Fashion Victim—Avoid the built-in obsolescence of trendy fashion statements by building your wardrobe around essential basics that won't date. Look for garments that are well made; they will both last longer and look better.

Washday Basics—Gentle laundering is a time-honored way of extending wardrobe life. Clothes and linens will last longer, for example, if you minimize the amount of chlorine bleach used for whitening. (Try cutting bleach in half and adding ¼ cup washing soda or borax to maximize the bleach's effectiveness.) Prevent snags by closing zippers and hooks-and-eyes before loading garments in the washing machine.

When the weather permits, forgo the dryer and hang laundry

on a clothes line or rack. Line drying is generally gentler to garments than machine drying, which can be especially hard on elastic-containing garments like socks and underwear. Best of all, line drying conserves energy and cuts air pollution. To keep garments from sun-fading, turn them inside-out.

A Stitch in Time Saves Nine—Stitch up rips and raveling seams promptly—before they grow in size. Small flaws are far easier to fix than large ones. Ditto for stains, which should be treated as soon as possible to prevent them from becoming permanent.

If your mom didn't teach you how to mend, a good book on clothing repair can provide clear detailed instructions almost as good as any mother's. (See ***Helpful Resources.***)

If you haven't the time or would rather not tackle a mending task yourself, ask a dry cleaner. Many dry cleaners perform simple mends and alterations, replace lost buttons and failed zippers. (See ***Zippers.***)

Defuzzing Devices—Sweaters and sweatshirts afflicted with pilling and fuzz balls are easily restored to like-new condition with a defuzzing device. There are several simple and inexpensive defuzzing devices on the market; a curved fingernail scissors also works well. (See ***Defuzzing Devices.***)

Turning Collars—On men's dress shirts, the collar is usually the first part to show signs of serious wear. If you have a sewing machine, you can conceal this wear and virtually double the shirt's useful life by carefully removing the collar, turning it over, and stitching it back in place. If sewing isn't your forte, plenty of dry cleaning establishments will do it for you—for a mere fraction of the cost of a new shirt.

Last Rites—When clothes are too frayed and faded to wear, you can cut them up for cleaning rags. Slipped over the hand and worn like a mitt, old socks are unbeatable for dusting. Terry towels make versatile and absorbent rags for all sorts of miscellaneous clean-ups in the house and garage. For polishing furniture, old flannel pajamas and worn out sweatshirts readily outperform their disposable paper counterparts. Except when used to clean up something toxic or unsavory, most rags can be washed and re-used many times.

Using cloth rags keeps single-use paper towels out of landfill. (Paper towels cannot be recycled because they contain a special wet-strength additive.) Additionally, relying on rags spares the trees and energy required to manufacture paper towels. It also diminishes water pollution caused by the chlorine bleaching agents used to turn brown wood pulp into bright white paper towels.

(Also see: ***Formal Gowns, Children's Clothing, Defuzzing Devices,*** and ***Zippers.***)

HELPFUL RESOURCES

1. ***The Lands' End Book of Caring*** provides handy laundering tips to preserve the quality, comfort, and fit of favorite garments. Available free from Lands' End, Inc:, 1 Lands' End Lane, Dodgeville, WI 53595-0001; (800) 356-4444.
2. ***Singer® Clothing Care & Repair: Extending the Life of your Clothes*** (Cy DeCosse, 1985, $10.95). This book demonstrates quick and easy ways to mend such commonly occurring problems as split seams, broken zippers, rips, tears, snags and stains. Available from **Cy DeCosse, Inc.,** 5900 Green Oak Dr., Minnetonka, MN 55343; (800) 328-3895.
3. Jan Saunders, ***Wardrobe Quick-Fixes: How to Update, Alter, Embellish, and Repair Your Clothes*** (Chilton, 1994, $14.95). Illustrated advice on how to fix stuck zippers and ripped out hems; how to shorten or lengthen, tighten or loosen clothes that don't quite fit. Available from **Nancy's Notions,** 333 Beichl Ave., P.O. Box 683, Beaver Dam, WI 53916-0683; (800) 833-0690.
4. **Bottomless Closet** provides preowned interview attire to women who are trying to leave the welfare rolls. Once they land a job, the women receive additional separates, dresses, and shoes. Since

it started five years ago in Chicago, the Bottomless Closet concept has spread to other U.S. cities. The organization's executive director, Kathy Miller, welcomes inquiries from individuals interested in donating gently worn office wear or establishing a Bottomless Closet in their city. 445 N. Wells St., Suite 301, Chicago, IL 60610; (312) 527-9664.

COFFEE FILTERS

There Are Safe, Re-usable Alternatives—Coffee filters made with paper whitened by chlorine bleach may contain trace amounts of the highly toxic synthetic chemical dioxin. According to a study performed by research chemists at Wright State University in Ohio, about half the dioxin in paper filters actually leaches into freshly brewed coffee.

Fortunately, there are several types of re-usable filters that won't pollute your coffee. These filters will preserve trees and decrease solid waste. And, over the long run, they will cost you far less than paper filters.

Cloth Filters—Filters made of unbleached cotton muslin may be re-used daily for up to two years. After each use, cloth filters should be rinsed thoroughly with plain hot water. (No soap, please, because it will leave a "yucky" tasting residue that flavors the coffee and is just about impossible to get rid of.)

To freshen a cotton coffee filter, soak it occasionally in a solution of baking soda and boiling water (2 cups water, 1 tablespoon baking soda). Rinse thoroughly before the next use.

Cloth filters are available in a range of sizes and shapes. Look for them in coffee specialty stores. (See ***Helpful Resources***.)

Gold-Plated Filters—Many connoisseurs prefer gold-plated filters to paper filters because the gold filter's fine mesh allows more of the flavorful coffee substance to pass through into the brew. Unlike paper filters, which can impart an undesirable flavor, gold does not affect the taste. Designed to last many years, gold-plated coffee filters are easy to clean. A thorough rinse under hot water is usually sufficient.

Gold-plated filters cost \$15–\$20. Look for them in department stores and coffee specialty shops.

French Press Type Coffee Makers—This system utilizes a plunger with a permanent mesh screen to filter the grounds from the brew. It produces a thick, heavy bodied coffee and is Thanksgiving Coffee roastmaster Paul Katzeff's favorite way to brew. He says it produces a cup of coffee that is "closest to the way professional tasters 'cup' coffees."

Look for French press coffee makers in the housewares section of department stores and in coffee specialty shops, or purchase direct from a mail order distributor. (See ***Helpful Resources.***)

HELPFUL RESOURCES

1. French press type coffee makers may be ordered directly from **Thanksgiving Coffee Company,** P.O. Box 1918, Fort Bragg, CA 95437-1918; (800) 648-6491.
2. Inexpensive cotton cloth filters (basket or cone type) may be ordered directly from **Clothcrafters, Inc.,** P.O. Box 176, Elkhart Lake, WI 53020; (800) 876-2009.

COSTUMES

For the Fun of It—Why buy a cheap throwaway Halloween costume when you can rent a fantastic re-usable one from a theatrical supply shop or a vintage clothing store? Better yet, create your own costume *for free* with materials scrounged from your closets, basement, garage, or—heaven forbid—the trash can or recycle bin.

Let the available materials inspire your imagination and go wild. Brainstorm with a friend. Improvise, invent, extemporize, and fake it. A touch of ingenuity, a few bits of colored paper (cut from magazines), and a dash of dime store glitter can turn a lowly egg carton into a jeweled crown fit for a princess.

If your creative juices need priming, let the local public library be your Halloween headquarters; borrow a book on costumes.

If brilliant ideas play hard to get or you run out of time, you can always rent a costume from a party equipment rental outlet, costume shop, theatrical supply company, or vintage clothing store. But the costumes that win accolades and prizes are the crazy one-of-a-kind creations made by re-using materials already on hand.

Share the Wealth—Don't forget that re-using channels goods to many users. So if you have a splendid Elizabethan gown, an aging Santa suit, or a fine vintage outfit that is only tempting the moths in your attic, consider offering it to a local drama group, theatrical supply company, or vintage clothing store. They might even pay you something for it. Or if you donate it to a nonprofit group, you could take home a tax deduction. (See Appendix A, ***Donate Instead of Dump.***)

CRAFT MATERIALS FOR KIDS

Letting the Genie out of the Bottle—What do you get when you turn thirty children loose in a room filled with clean industrial doodads and discards, fabric remnants and ribbons, paper scraps, art paints, and crayons?

If you assure the kids that there is no "right" or "wrong" way to use the materials at hand and tell them they can do whatever they want, you'll witness an explosion of creative chaos. The kids will express themselves freely, enjoy themselves immensely, and create the darndest things.

The Doodad Dump is one of the most popular areas at the Children's Discovery Museum in San Jose, California. Here, armed with hot glue guns that have been modified for safety, children transform industrial discards into truly unique creations.

Teachers bring classes here on field trips, but families and individual children are equally welcome. Parents and teachers may also take home grocery sacks filled with discards for $5 apiece.

One Saturday each month, teachers may visit an off-site warehouse where, in addition to "idea" sheets with art project suggestions, they can obtain a mind-boggling variety of industrial scraps at a cost of about $2 to $4 per grocery bag—a real bargain when you consider the high cost of most classroom craft materials. Naturally, purchasers are encouraged to bring their own previously used grocery bags for hauling goodies home.

The Doodad Dump is just one of a growing number of Teacher Resource Centers (TRCs), a new type of nonprofit organization developed specifically to provide teachers, parents, scout leaders, artists, and community organizations with inexpensive art materials and creative play things. Because children learn best by doing, many TRCs offer teacher workshops that demonstrate ways to re-use discarded materials in math, science, and art projects.

A Businessman's Trash Is a Teacher's Treasure—Most of the art materials available at TRCs are industrial scraps and byproducts donated by local businesses. TRCs welcome safe, clean wastes of every description, including paper products, office supplies, art supplies, fabric remnants, spools of yarn, and miscellaneous plastic doodads.

In communities that don't have a TRC, teachers and schools can establish their own exchange relationships with businesses in the area. When teachers try their hand at soliciting local industries, they are often pleasantly surprised at how eager many companies are to support their community's schools while cutting their own disposal costs. Donations should be carefully screened, of course, for safety and usability. A school that accepts materials it cannot use may find itself burdened with unanticipated disposal expenses.

HELPFUL RESOURCES

1. Interested in how kids develop creativity and problem-solving skills with clean industrial discards? Dr. Walter Drew's booklet, ***Recycled Materials—Tools for Invention and Self Discovery*** is available for $5 from **The Institute for Self-Active Education,** P.O. Box 1741, Boston, MA 02205; (617) 635-8284.
2. Want to know if there is a Teacher Resource Center in your area? Contact your city or county recycling department.
3. Can't find a TRC in your area? Order an inexpensive assortment of industrial odds and ends through the mail from the **Boston Schools Recycling Center,** P.O. Box 1741, Boston, MA 02205; (617) 635-8284.
4. Want to set up a TRC in your community? **Materials for the Arts in New York City** publishes a free step-by-step handbook on how to secure sponsors, obtain funding, and contact businesses that might be willing to donate materials. The guide is available from Materials for the Arts in New York City, 410 W. 16th St., New York, NY 10011; (212) 255-5924.

DEFUZZING DEVICES

With the help of a defuzzing device, which removes unsightly pills and fuzz balls from sweaters and napped fabrics, you can restore a garment's original smooth appearance and extend its useful life considerably.

My favorite defuzzing device is a curved fingernail scissors. The beauty of the curved scissors is that it's multi-functional. You can use it for trimming fingernails *and* for defuzzing sweaters. If you've already got such a scissors in your manicure kit, you needn't buy anything else. When the blades get dull, you can take the scissors to a knife sharpener and have it sharpened.

Whenever fuzz balls appear on a sweater, I clip them off with my curved scissors, taking care to clip only the pills and not the sweater. Within minutes, the sweater looks brand new again. The fingernail scissors works well on larger defuzzing tasks, too. About once a year when I really want to impress the in-laws, I clip the pills off the upholstery on the living room sofa.

For gadget lovers, there is a variety of defuzzing devices on the market. Before experimenting with any of these devices, be sure to read the manufacturer's instructions carefully; then test the device on a small and inconspicuous section of the fabric.

1. The **d-fuzz-it®** is an inexpensive (about $2) handheld tool that you brush quickly and lightly over the fabric surface. It can be used on sweaters of wool, cashmere, or acrylic as well as on coats and blankets. The d-fuzz-it separates the pills from the fabric. Then you brush the garment off with your hand or a damp washcloth.
2. About the size of a bar of soap, the **Sweater Stone®** has a pumice-like surface that lifts pills from a variety of fabrics including suede, wool, cotton, and even some upholstery. I have found that the Sweater Stone works wonders on cotton/polyester sweatshirts, which have a lamentable tendency to develop zillions of tiny pills. Since this device sheds small grains

of abrasive stone that look like they could wreak havoc on hardwood floors, I recommend shaking garments off outdoors after defuzzing. The Sweater Stone costs about $8.

3. According to the manufacturer, Remington's handheld battery-operated **Fuzz-Away®** removes pills and fuzz from clothing, draperies, upholstery, and even carpet. I expected this electrically powered gadget to speed up the defuzzing task dramatically, but found it requires just as much time and patience as the human-powered devices. The Fuzz-Away costs about $7, plus the cost of batteries.

HELPFUL RESOURCES

1. The **d-fuzz-it®** is available in sewing and fabric stores, or by mail from **Nancy's Notions** at 333 Beichl Ave., P.O. Box 683, Beaver Dam, WI 53916-0683; (800) 833-0690.
2. The **Sweater Stone®** can be ordered from **The Vermont Country Store** at P.O. Box 3000, Manchester Center, VT 05255; (802) 362-2400.
3. **Remington's Fuzz-Away®** can be purchased in the small appliance section of large drugstores and super stores.

DIAPERS

The Siren Call of Convenience—85 percent of American parents diaper their babes in disposables. Why? Because they are convenient.

Fortunately for the environment, disposable diapers do not have a monopoly on convenience. Cloth re-usables, laundered by a professional service and delivered to your front door once a week, can be every bit as convenient as cellulose disposables. You don't have to lug huge packages of disposables home from the grocery store. And you don't have to rinse, soak, or flush soiled diapers. The diaper service takes care of all that.

Professional diaper services have diaper washing down to a science. They know exactly how hot the wash water should be and how much bleach is needed to destroy all known microorganisms. They know which detergents are least likely to leave rash-causing residues, and how to inhibit diaper rash by controlling the pH level of the finished diaper.

New designs in washable diaper covers have eliminated the need for pins, making diaper changes quick and easy. These covers are fastened with **Velcro®** tabs or snaps. All you have to do is lay the folded diaper in the cover and wrap it around the baby.

If you launder at home, there are even more handy new diaper designs. There are contoured diapers that don't require folding. And there are all-in-one diapers that combine a cloth diaper with a leak-preventing lining and Velcro fasteners. These time- and sanity-saving variations on the old-fashioned cotton diapers may be purchased from specialty baby stores, department stores, or some kids' wear catalogs. (See ***Helpful Resources.***)

The Cost Factor—If you don't take the environmental costs of disposables into account, the price of single-use throwaways and professionally laundered diapers is fairly comparable—about $13 per week for a newborn baby. Home laundering is quite a bit cheaper, though if you opt for the convenience of all-in-one diapers the cost of your initial investment will be somewhat higher than if you stick with traditional rectangles.

Child Care Center Concerns—Most day care centers ask parents to supply disposable diapers, but some are glad to use cloth as long as parents pick up dirty diapers daily and provide an adequate supply of diapers, pins, and plastic pants, along with an airtight container for storing used diapers.

A recent study at the Vanderbilt University Medical Center found that diaper type makes no difference in the presence of diarrhea-causing bacteria at day care centers. "The key to wellness in child care centers," the study concluded, "is handwashing and hygiene practices, not diaper type."

Congratulations—Parents who opt for re-usable diapers deserve high praise for their environmentally considerate ways. Their efforts are helping to create a more habitable world for all of us. It has been estimated that each baby who spends the first two years of life in washable diapers conserves about twenty trees; trees that would otherwise be felled for the cellulose needed to manufacture two years' worth of disposables. The baby swaddled in washables also prevents about a ton of solid waste from burdening local landfills.

Gift Idea—A month of diaper service makes an excellent shower gift. Better yet, combine forces with others and give several months of service.

HELPFUL RESOURCES

Pinless diaper covers, contoured no-fold diapers, and all-in-one diapers may be purchased in baby and department stores or by mail from:

Biobottoms, P.O. Box 6009, Petaluma, CA 94955-6009; (800) 766-1254.

The Natural Baby Catalog, 816 Silvia St., 800 B-S, Trenton, NJ 08628-3299; (800) 388-2229.

Babyworks, 11725 N.W. West Rd., Portland, OR 97229; (800) 422-2910 or (503) 645-4349.

DOG and CAT SUPPLIES

The Humane Society Practices Re-use—Everybody knows you can get a preowned cat or dog at the Humane Society's animal shelter. But did you know that many animal shelters also accept —and re-use—previously owned pet care items? The Society for the Prevention of Cruelty to Animals (SPCA) in San Francisco, for example, welcomes a variety of used pet items, including crates and carriers, metal food and water bowls (no plastic, please), and doggie sweaters. Also flea combs, dog collars and leashes, provided they're in good condition.

Throw in the Towel—Believe it or not, old towels are a very popular pet care item. I've noticed signs soliciting old towel donations in waiting rooms at both the veterinarian's office and the kennel. The local Humane Society welcomes old towels too. They use them for bedding and for drying just-bathed dogs.

Re-using the News—Plenty of cost-conscious cat owners have been known to hand-shred newspapers into thin strips for use as cat box filler. It's low tech, but it works. And it's so cheap you can afford to change it every day.

Tennis Balls, Anyone?—Some kennels like to receive funky old tennis balls that have lost their bounce. They use them for exercising the dogs. (But please don't let your dog use tennis balls as a chew toy. Tennis ball innards are bad for Bowser's digestion.)

DRAPERY

Even though you may be sick and tired of looking at them, your old draperies may be just what the decorator ordered for someone else's home. Provided they're in decent condition (not too sun-ravaged or stained), you can channel your old window coverings to a new home by offering them to a charity thrift store such as Goodwill or Salvation Army. Or call a home rehabilitation program such as Habitat for Humanity or Christmas in April to ask if they can use them. (See ***Building Materials.***)

Who knows, a theatrical group may find your drapes to be just what the set designer specified. Look for theater groups in the Yellow Pages under "Theaters-Live"; or contact high school and college drama departments. Some theatrical equipment and supply companies may also be interested in your drapes.

Drapery too decrepit to grace a window can serve in other, humbler ways—especially if made of fabrics that are washable and absorbent. If the fabric is still sturdy enough, I'll use old curtains to make washable covers for the styrofoam beds my dogs sleep on. Many launderings later, I'll snip the covers into rags for cleaning and house painting projects.

DRY CLEANING BAGS

Whenever I come home from the dry cleaners with an armload of freshly laundered shirts, I remove the plastic polybag that covers them and . . . true confessions . . . I stuff it in the trash. I am hardly alone in this, of course, for it has been estimated that more than one billion such plastic film dry cleaning bags get thrown away each year.

A Better Way—Doesn't anybody care about all this plastic waste? Yes, somebody does care. A company called Safety-Kleen, which cleans and recycles the solvents used by many dry cleaners across the country, has initiated a WE CARE program in which participating dry cleaners sell two re-usable fabric garment bags to each customer. The dry cleaner keeps one bag at his shop, and you take your clean clothes home in the second bag. Once home, the second bag serves as a duffle bag for dirty clothes. When it's full, you take it to the cleaners and exchange it for a clean bag containing clean clothes.

Voila! A re-use cycle that is every bit as convenient as stuffing plastic bags in the trash.

Cleaners who participate in Safety-Kleen's re-usable garment bag program usually display the WE CARE symbol in the window. Ask your dry cleaner about this program. If they don't participate, encourage them to find out about the program. (See ***Helpful Resources***.)

Good News—In 1991, New York City's Neighborhood Cleaners Association, in partnership with the city's Department of Sanitation, launched a waste prevention campaign that encourages customers to use re-usable garment bags, return wire hangers for re-use, and return plastic polybags for recycling.

If this program wins the cooperation of a majority of member dry cleaners, the Sanitation Department estimates that the savings to taxpayers could easily exceed one million dollars annually in preserved disposal capacity and avoided collection costs.

A 1995 progress report on the program notes that dry cleaners participating in the partnership are achieving substantial cost savings. In the East Village, the Chris French Cleaners persuaded 440 regular customers to adopt re-usable garment bags. As a result, that cleaner's orders for plastic film bags have decreased by 90 rolls per year—saving the cleaner $2,250 annually and keeping nearly one ton of polybags out of the city's waste stream. (Also see ***Wire Clothes Hangers.***)

HELPFUL RESOURCES

1. Would you like to encourage your city's sanitation department to join forces with local dry cleaners in promoting re-usable garment bags? To learn more about New York City's partnership with the **Neighborhood Cleaners Association,** contact the **Bureau of Waste Prevention, Reuse and Recycling Dept.,** 44 Beaver St., 6th floor, New York, NY 10004; (212) 837-8183.
2. Dry cleaners interested in re-usable garment bags may contact these companies:

    Safety-Kleen, 777 Big Timber Road, Elgin, IL 60123; (800) 669-5740, ext. 2609.

    Laundry Bags Unlimited, 310 Paterson Plank Rd., Carlstadt, NJ 07072-2306; (800) 678-4LBU.

ENVELOPES

"Thrift Is a Great Revenue"—Companies that manage to survive in today's super competitive business environment know—as did Cicero more than two thousand years ago—that "thrift is a great revenue." That's why re-using heavy-duty manila envelopes for interoffice mail is a well-established tradition in the world of commerce.

Similarly, when companies mail out bills in re-usable two-way envelopes that customers can re-use to send in payments, they don't do it just to save trees and landfill space. They do it because it's good for their bottom line—one two-way envelope costs less than two one-way envelopes. The state of Illinois, for example, has been sending out license plate renewal applications in two-way envelopes since 1984. In 1990, this strategy saved the state $57,000. Individual householders can reap their own personal savings from envelope re-use.

Preaddressed Reply Envelopes—Junk mail solicitations often contain preaddressed reply envelopes, which are handy for sorting and storing grocery coupons, taking phone messages, or scribbling "to do" lists. When preparing for a trip, I use individual reply envelopes to pack each daily dose of vitamin pills.

When appearances are not important, you can also use reply envelopes to send personal mail or pay bills. Just cover the address with a mailing label and write or type in your own addressee. Be sure to obliterate with a magic marker any bar code address (printed near the envelope's bottom edge); otherwise the Post Office computers will automatically direct your missive to the original preprinted addressee.

If you can get them, it's best to use the old-fashioned lick-and-stick type of labels that have a water-soluble gum backing. Envelopes with lick-and-stick labels can be recycled; envelopes with self-adhesive labels can *not* be recycled, because the insoluble adhesive tends to clog paper-making equipment. (See ***Helpful Resources.***)

Oversized Envelopes—I frequently receive printed materials in 9x12-inch manila envelopes. Whenever possible, I try to re-use these envelopes for my own sending purposes. But this only works if the original sender has been environmentally considerate and closed the envelope with the brass clasp or with easy-to-remove transparent tape. If the gummed flap has been sealed, the envelope is not usually re-usable.

HELPFUL RESOURCES

Old-fashioned lick-and-stick address labels are hard to find at stationery stores, but you can order them from the **Oakland Recycling Association**. Made with unbleached paper, these handsome 4x5-inch labels are large enough to cover the address on a previously used envelope. The cost is $3.50 for 100, $10 for 300. Send your request, along with check or money order, to Oakland Recycling Association, P.O. Box 114921, Oakland, CA 94611-0492; (510) 444-8633.

EYEGLASSES

Poor eyesight can be devastating. It can force adults out of work and children out of school. According to the World Health Organization, one in every four people needs corrective lenses, but most have no hope of affording glasses.

To serve these people, the Lions Club International collects old, unwanted glasses and distributes them free of charge to needy people in developing nations, where a pair of spectacles can cost as much as a month's wages.

Anyone with a pair of old prescription eyeglasses collecting dust in a closet or drawer can make a valuable contribution to this re-use program, which also welcomes ordinary nonprescription sunglasses. All year round, and especially during the month of May, "Lions Recycle for Sight" collection bins are displayed in many places of business throughout the United States and Canada. Unwanted specs may also be dropped off at any one of more than 650 LensCrafters outlets throughout the United States.

The collected spectacles are forwarded to one of five regional eyeglass recycling centers, where volunteers clean and repair, sort and categorize them according to prescription. The eyeglasses are then distributed to optical missions in developing countries sponsored by Lions Clubs, LensCrafters, and other service groups.

HELPFUL RESOURCES

1. For the location of the nearest **Lions Club** eyeglass collection bin, call a local Lions Club office. Or contact the **International Association of Lions Clubs** at 300 22nd St., Oak Brook, IL 60521-8842; (708) 571-5466.
2. For the location of the nearest **LensCrafters,** call 800-522-LENS.

FLORIST VASES

Plenty of florists are delighted to take back vases and baskets for re-use. A phone call or two should be all it takes to locate a nearby florist who welcomes such containers.

Most florists don't re-use floral foam (the green stuff in the bottom of the vase that holds plant stalks in place). But it's such handy stuff that you might want to try re-using it yourself for homemade bouquets.

FORMAL GOWNS

Until a few years ago, purchasing dresses and gowns for formal occasions could—and often did—wreak havoc on a woman's budget. A man could rent a tuxedo for about $75, but a woman usually had to lay out several hundred dollars for suitable attire which, thanks to the fickleness of fashion, she might wear only once.

Fortunately, this situation has begun to change, and now women, too, can rent instead of buy.

Rent-A-Gown—Bridal and formal wear shops that rent designer gowns and dresses—for about one-third of the retail purchase price—offer welcome relief to recession-pinched pocketbooks. These shops rent high quality designer gowns, dresses, and cocktail suits for every kind of special occasion including weddings, proms, balls, cocktail events, pageants, and cruises.

When I spoke with the folks at "Formal Rendezvous," a gown rental shop a few miles south of San Francisco, I learned that several of the women who attended President Clinton's inaugural balls had gone there to lease formal frocks for the special occasion.

Gown rental outlets stock the latest fashions as well as classic designs in a full range of standard sizes. A few shops carry maternity and petite-sized gowns as well. Many outlets will make simple alterations for free, while some charge extra.

Hygiene is probably the most frequently raised objection to gown rental. Lots of women are frankly squeamish about donning a dress that someone else probably perspired in. Gown rental stores assure me, however, that—like tuxedos—all gowns are professionally dry-cleaned after each rental.

Wedding Bells—Although lots of women still want to tie the knot in a brand new virgin-white gown, plenty of pragmatic brides do otherwise. If you are reluctant to blow a small fortune on a single-use garment, you owe it to yourself to investigate the offerings in consignment shops, where breathtaking designer gowns may be had for less than half the original price. Some gown rental shops will also sell items from their inventory at reduced prices. Or let serendipity be your guide and check out the bridal gowns offered for sale in newspaper classified ads, flea market bulletins, and company newsletters.

Instead of a white costume you'll only wear once, consider getting married in an elegant dress you can wear again. In England's Victorian era, this was very much the custom. In those days, a bride's wedding dress varied only slightly from everyday fashion, and she wore it on special occasions for many years.

HELPFUL RESOURCES

Look for gown rental shops in the Yellow Pages under "Formal Wear—Rental and Sales" or "Bridal Shops."

FURNACE FILTERS

The vast majority of furnace filters sold in hardware stores today are short-lived disposables. If you rely on these single-use products and change your furnace filter as often as recommended, during seasons of heavy use you'll be purchasing a new one every month or two—and stuffing the old one in your garbage can.

Although not as widely available as disposables, re-usable furnace filters are certainly not difficult to obtain. Many of the newer re-usable filters now on the market are far more effective at trapping tiny airborne particles than ordinary fiberglass disposables, which catch some of the largest particles but retain relatively few of the smaller ones.

Usually made of synthetic fibers such as polyester and polyethylene, the new re-usables rely on "static cling" to attract and hold microparticles. When air flows through these filters, the synthetic materials rub together, and the resultant friction creates an electrostatic charge capable of capturing more than 90 percent of airborne particles.

Re-usable fiber filters can usually be cleaned by rinsing them with water. Remove the filter from the furnace and follow the manufacturer's instructions. Always allow filters to dry thoroughly before replacing in the furnace. If you invest in two washable filters, you will have one to install while the other dries.

Calculate the Savings—Initially, re-usable furnace filters cost more than standard throwaways. Over time, however, two re-usables can spare you the cost of many more single-use filters. You'll be conserving nature's resources too—including the energy and materials needed to manufacture dozens of fiberglass filters and the fuel required to transport them from the factory to the store.

By cleaning furnace filters frequently you can save still more energy and could even prevent premature blower failure. Dirty filters demand more energy than clean ones, because the fan must work harder (and consume more electricity) in order to force air through a filter clogged with dirt.

HELPFUL RESOURCES

Look for washable filters in hardware stores and home centers, or purchase them by mail.

The **Web®** is a washable two-stage electrostatic filter made of polyester and polyethylene. This filter comes in sizes that fit all standard furnace and air-conditioning systems, costs about $25, and carries a three-year warranty. It is sold in hardware stores and home centers nationwide. For additional product information or the name of a local outlet, contact WEB Products, Inc., 11 Lincoln St., Kansas City, KS 66103; (800) 875-3212.

Air Magnet® is a 5-layer electrostatic filter of polypropylene and polyurethane. It is somewhat more expensive (about $100), but very long-lived and comes with a lifetime warranty. For the name of a local distributor, contact Air Magnet, 2107 Lawn Ave., Cincinnati, OH 45212-9899; (800) 743-9991 or (513) 731-9991.

Allergy Resources, Inc., sells high efficiency washable furnace filters. To request a catalog: P.O. Box 444, Guffey, CO 80820; (800) USE-FLAX or (719) 689-2969.

FURNITURE

Rescuing old furniture from the trash heap is a healing activity, whether it's a comfy old armchair you've lived with for years or a homely flea market find.

When a human being breathes new life into an old piece of furniture, a tree is spared from the chain saw and energy is saved—energy that would otherwise be needed to chop down that tree, mill it, manufacture it into tables and chairs, and transport the finished goods to market. Conserving that energy translates into less air pollution, fewer greenhouse gas emissions, and less acid rain. And this, of course, means healthier people and forests.

When we do something that has a restorative effect on the earth, we also heal an important part of ourselves that lies deep inside the human psyche where each of us is sympathetically bonded to the biosphere that gives us life.

Refinishing and Repairing—Sometimes a gentle cleaning is all that's needed to bring out the warm patina of life in a well-used piece of furniture. With good instructions and the right supplies, almost anyone can do a bit of regluing, make simple repairs, and smooth away minor blemishes.

Most book stores and libraries are well stocked with how-to books on furniture refinishing and repairing. Keep an eye out for helpful books at library fund-raiser sales and used-book stores. Often written by professional furniture restorers, these books can guide you in diagnosing problems and familiarize you with the arsenal of techniques and products used by the pros. (For best results, always follow manufacturers' instructions carefully.)

Reupholstering is Planet-Friendly—Have you a sofa that is badly faded or unreclaimably dirty? Provided it was moderately well built in the first place and remains structurally sound, your sofa may be a top candidate for reupholstering.

By reupholstering you'll save quite a chunk of Mother Nature's precious resources, including the wood frame, the metal springs, as well as some padding and stuffing materials.

Don't expect to save a bundle by reupholstering, though, because you could probably buy an inexpensive new sofa for less than the cost of a professional reupholstering. The snag with low-priced sofas is that few of them are sturdy enough to withstand reupholstering, so what you are purchasing is essentially a "disposable" sofa that may not last more than a decade before becoming a landfill behemoth.

To determine if a piece of furniture is sturdy enough to warrant reupholstering, you'll have to peek under the surface and check out the frame. To do this, turn the piece over and remove a portion of the cambric dustcatcher underneath.

A frame made of hardwood at least one inch thick that is doweled and glued together is far more likely to hold up through at least one round of reupholstering than a frame made of plywood, fiberboard or soft pinewood that's stapled together. Other signs of quality construction include the use of reinforcement blocks at the corners and hand-tied coil springs.

Recovering the cushion on a wooden chair is a relatively simple task that any moderately determined person can tackle with confidence. Reupholstering sofas and covered chairs, however, is usually a job for a professional. Nevertheless, for intrepid do-it-yourselfers with solid sewing skills, there is a substantial body of literature on the art of upholstering. Who knows, you might discover a new career.

Slipcovers—Easier and less expensive than reupholstering, slipcovers are an elegant way to revive threadbare furniture or to protect it from heavy use. Removable, fitted cloth covers for armchairs and sofas, slipcovers are especially practical in families with children because they protect furniture and can be made with fabrics that are washable. When a slipcover is soiled, it can be easily unzipped for laundering or dry cleaning.

Ready-made slipcovers can be purchased in department

stores, but the best-fitting slipcovers are custom made. If you are handy with a sewing machine, you can save money by creating your own custom covers.

Maintenance Tips—Over the long haul, taking good care of your furniture is a whole lot easier than repairing and refinishing. Help your furniture to age gracefully by following these basic precautions:

- Protect wood and upholstered furniture against drying and fading by drawing drapes against intense afternoon sun;
- Avoid placing furniture in front of radiators, heating registers, and fireplaces;
- Sofa cushions last longer and wear more evenly when rotated and turned periodically;
- Regular vacuuming with an appropriate attachment keeps dust particles from getting ground into upholstery fabrics.

HELPFUL RESOURCES

1. ***Doing Up Old Junk: How to Revamp Shabby Furnishings with Style*** by Joanne Jones (Merehurst, 1994, $19.95). Full color photographs demonstrate the equipment and techniques needed to fix up old junk with flair and imagination.
2. ***Reupholstering at Home: A Do-It-Yourself Manual for Turning Old Furniture into New Showpieces*** by Peter Nesovich. (Crown, 1988, $9.95). Written for the novice, this manual contains step-by-step instructions and hundreds of photographs and illustrations.
3. ***Care and Repair of Furniture*** by Albert Jackson and David Dey (Taunton, 1994, $27.95). Step-by-step directions, full color illustrations and dramatic explosion views show you how to diagnose damage, match finishes, make strong, virtually invisible repairs for many additional years of use.
4. ***Slipcover Magic*** by Dorothea Hare and Ron Caralissen (Chilton, 1995, $21.95).

GARDEN

Re-using the Earth—If attentively maintained and repaired as needed, an automobile or washing machine may provide several decades of reliable service. But a patch of earth that is well-maintained and regularly replenished can serve virtually forever.

Knowledgeable gardeners recognize that the best way to sustain a healthy garden over time is to provide regular infusions of properly composted organic matter. Dug into the soil or spread as a mulch, compost promotes healthy soils by introducing nutrients, minerals, beneficial microorganisms, and organic matter essential for building healthy, insect- and disease-resistant plants. According to Mike McGrath, editor of *Organic Gardening* magazine, "one inch of compost on top of the soil, not mixed in, is more effective than any fungicide." (But keep mulch several inches away from tree trunks.)

When dug into the soil, the humusy organic matter in compost improves drainage and makes oxygen more available to roots. When the organic matter bonds to certain minerals and micronutrients in the soil, it makes them more available to plants. Some of the microorganisms in compost take nitrogen from the air and make it accessible to plants. Still other microorganisms produce antibiotics that protect plants from diseases.

Team Up with Nature—You can buy bagged compost at any nursery, but why bother when Mother Nature makes it so miraculously easy and inexpensive to produce compost right in your own backyard?

Instead of stuffing garden trimmings in the trash can, just heave them, along with fallen leaves and lawn clippings, on a compost heap. Spread several shovels full of cow manure on the pile, moisten it with a hose, and let it rot. Each time you add several inches of garden waste, be sure to top it off with an aromatic dose of manure. Experts recommend a ratio of about three parts plant debris to one part manure.

Using special compost containers and occasionally aerating the pile by turning the contents with a pitch fork will speed up the rotting process, but it's not essential. If you don't bother to turn the pile, in about a year you'll have a luscious pile of crumbly black humus that smells like the forest floor. Fact is, on a small scale, backyard composting mimics the way plant materials naturally decompose on a forest floor.

Composting Makes $ Sense—It has been estimated that by directing garden waste to a compost heap, it is possible to reduce household waste going to landfills by 25 to 35 percent. Home composting often enables a household to cut back from two garbage cans to one, making it possible to save well over a hundred dollars a year in disposal costs. In the community where I live, we pay about $16 per month for one can; $32 for two cans. So eliminating the second can saves us $192 a year. That's nearly $2,000 in a decade!

For additional savings, don't forget the money you won't have to spend on bagged compost at the nursery. The cost of the manure you'll need to get your compost heap cooking is a drop in the bucket compared to what you'd pay for commercial soil amendments.

HELPFUL RESOURCES

Compost aficionados call the simple composting method I've described here a "passive pile." If you would like to learn about more "active" styles of composting, there are plenty of books on the subject. Among my favorites are:

Easy Composting (Ortho Books, 1992, $9.95)

Let it Rot: The Gardener's Guide to Composting by Stu Campbell (Storey Comm, Inc., 1990, $8.95)

GIFTS

Help others conserve energy and resources by selecting gifts that are re-usable or environment-friendly. The best gifts add more to the world than they consume:

1. A battery charger with rechargeable batteries makes a great gift for a child who enjoys battery-operated toys—or for an adult who uses a portable tape recorder, radio, beeper, or dictaphone. (See ***Batteries.***)
2. For soon-to-be parents, how about a gift certificate for a month or two of diaper service? Hopefully, after a month, they'll be hooked on this convenient and environmentally considerate method of diapering. (See ***Diapers.***)
3. Just about everyone loves to receive good food. High quality fruit, nuts, preserves, or breads will almost always be appreciated.
4. Plants are ideal gifts for a host or hostess.
5. Who wouldn't appreciate biodegradable lotions, soaps, or shampoos in refillable containers?
6. Canvas or string shopping bags make useful gifts.
7. Colorful cotton kitchen towels are a welcome addition to anyone's kitchen. They can double as gift wrap too.
8. When selected with care, secondhand toys and sports equipment make first rate gifts for youngsters. Most of the upscale resale boutiques specializing in toys carry high quality items that have barely been used; some even have the original packaging and instructions. Garage sale finds may need a bit of sprucing up. For tips on revitalizing previously cherished playthings, see ***Toys.***
9. Psychological studies show that social relations and leisure time are primary sources of human fulfillment. So how about giving your significant other or anyone else you love a gift of leisure time together? Perhaps an afternoon outing or a leisurely lunch together, a visit to a favorite museum, concert tickets, whatever it is you both enjoy.

10. A good friend might welcome your gift of an evening's babysitting or a healthy home-cooked meal.
11. Are you handy at fixing things? If a good friend has something that's broken but still useful or valuable, why not make a gift of your skills by offering to repair it?

Unwanted Gifts—When you receive a gift you cannot use or just plain don't want, put it aside with a note indicating who gave it to you. Later on, you can give it to someone else for whom it may be more appropriate. The note will keep you from inadvertently—oops!—giving it back to the person who gave it to you.

White Elephant Parties—In my husband's office, the annual Christmas party is a re-use event. Each person brings a "white elephant," usually a present he or she once received but never cared to use. These gift exchanges inevitably bring lots of laughs; it's fun to see what unprized items others are eager to unload on their coworkers.

Over the years, my husband has brought home from these parties an odd assortment of items: a Chinese foot massager, a cookbook, a promotional coffee cup emblazoned with a company logo, and a supremely ugly tie clip. Don't be surprised if that tie clip turns up again at the next Christmas party.

GIFT WRAP

By wrapping gifts in re-usable materials you may already have on hand, you can unclutter your closets, have a little creative fun, and even save a tree!

Colorful scarves, cloth napkins and even dish towels make excellent gift wrap, as do fabric bags tied up with a pretty ribbon, and all are eminently re-usable. Or it's easy to make brightly colored fabric squares using material left over from sewing projects or remnants bought inexpensively on sale. Just cut the cloth to size and run a zigzag stitch around the edges to prevent raveling. Use it as you would regular wrapping paper, secured with a ribbon. The gift recipient can press out the creases with a steam iron and the wrapping is ready to re-use.

Something elegant but simple, particularly for gifts that consist of several small items, is a basket. Arrange the gifts attractively and add a bow or other ornament. The basket is instantly ready to re-use for any purpose, including future gift giving.

Even if you prefer to wrap your gifts in paper, there are several re-usable alternatives. One effortless way to present your gift is in a gift box. Perhaps you've got a couple of used gift boxes stashed in a closet, just waiting for an opportunity to be re-used. Drop the gift in the box, tie a ribbon around it and it's ready to go, no wrapping required. The same is true for gift bags made of sturdy paper that are widely available in drug and stationery stores. I've got one that's been in service for four seasons, and it's still in like-new condition.

For a specialized, yet practical touch, why not delight a child with a gift wrapped in the Sunday funny papers. And wouldn't Grandma love to receive a gift covered in her grandchild's artwork?

Speaking of Grandma, my own grandmother was a genius with previously used wrapping papers, relying on strategically placed ribbons, bows, and gift cards to conceal creases and small tears. She usually left discarded ribbon for the cat to play with, but plenty of folks re-use ribbons too. I know of one serious waste reducer who stores her re-usable ribbons by winding them around a cardboard toilet paper tube!

If wrapping paper won't be re-used, set it aside for a rainy day when restless young children need something to do. They can use it for an art project.

GOLF BALLS

Because I'm always on the lookout for novel approaches to re-use, I was delighted to learn that golf ball retrieval is a flourishing business.

Second Chance, reputedly the largest golf ball retrieval company in the country, dredges 60 million balls a year from the waters of a thousand golf courses across the nation. Some retrievers wear wet suits and scuba diving gear to rake the lake bottoms for stray balls. Others utilize a mechanical device that goes under water and scoops them up. The reclaimed balls are cleaned and graded for resale through golf courses or mass merchandisers.

Resourceful golfers can scrounge plenty of balls for themselves. Sleuthing in the undergrowth just outside the fence surrounding the local public golf course, my husband regularly finds a bonanza of errant missiles. Thanks to this year-round Easter egg hunt, he never needs to buy golf balls.

If reconnoitering in the rough is not your style, you can purchase retrieved balls in pro shops and golf supply stores. The pro at a nearby public course tells me that his shop buys retrieved balls wholesale and resells them to golfers for $1 apiece—a bona fide bargain compared to $2.50 or more for new ones. The pro assures me that the retrieved balls he sells are in very good condition. Some have only been hit once, and all are clean and free of cuts or abrasions.

GREETING CARDS

Greeting cards are designed to be used once and tossed in the trash. But it doesn't have to be that way, because re-using greeting cards is both easy and convenient.

Cut the front off cards you receive and re-send them as postcards or thank-you notes. This only works, of course, if the back side hasn't already been written on. By the way, these front panels can also make good gift tags.

If you don't re-use greeting cards yourself, grant them a second life by sending them to a charitable organization that does re-use.

HELPFUL RESOURCES

1. **St. Jude's Ranch for Children** is a nonsectarian facility for homeless children. The organization re-uses old Christmas, birthday, get well, sympathy, and general cards. The children cut, trim, and paste the card fronts onto new card backs. From the sale of these "new" cards, the children earn pocket money and learn valuable lessons about the work ethic and the value of a dollar.

   Send card fronts only (this saves postage) to: St. Jude's Ranch for Children, 100 St. Jude's Way, P.O. Box 60100, Boulder City, NV 89006-0100.
2. **All Year Christmas Cheer** does not accept greeting cards directly, but can provide names and addresses of schools, orphanages, and missions that welcome used greeting cards of all kinds, including Easter, Valentine, Mother's Day, birthday, etc. For the name and address of nonprofits that will put your old greeting cards to practical use, send a stamped self-addressed envelope to All Year Christmas Cheer, 134 Pfeiffer St., San Francisco, CA 94133.

GROCERY BAGS

It has been estimated that Americans use about 40 billion disposable grocery bags every year. Stated this way, the problem seems enormous, even insurmountable. But never underestimate the little decisions of daily life. Small changes, when implemented by many people, can have an aggregate impact that boggles the mind. For example, if every household in the United States used one less disposable grocery sack per week, we'd conserve the energy, oil, and trees needed to manufacture more than five billion bags every year. We'd save a lot of landfill space too.

Re-usable alternatives to disposable bags are plentiful and varied. You can start by re-using grocery bags already collected from previous shopping trips. (A growing number of grocery stores offer refunds of up to 5 cents for each bag that customers re-use.) You can also purchase inexpensive canvas or string bags in supermarkets, natural food stores, and environmentally oriented "green" stores.

The most durable bags are made of washable cotton canvas. String bags and net bags are convenient because they can be easily folded and stowed in a purse or bookbag.

To develop the habit of bringing your own bag, keep one or two in the car so they'll be there when you need them. After unloading groceries, make it a habit to put bags back in the car or near the front door, so you don't leave home without them.

Now Here's a Bright Idea!—Quality Food Centers, a supermarket chain in Washington state, encourages shoppers to re-use grocery bags regularly. For every large bag that customers bring in and re-use, the store donates 3.5 cents to The Nature Conservancy, an organization dedicated to preserving endangered ecosystems. Over the last six years, by re-using their grocery bags, shoppers have raised over $225,000 to purchase endangered wildlands in the state of Washington.

HELPFUL RESOURCES

Would you like to encourage a local supermarket to promote grocery bag re-use while raising funds to protect endangered ecosystems in the state where you live? Contact **The Nature Conservancy,** 1815 North Lynn St., Arlington, VA 22209; (703) 841-5300.

HEARING AIDS

Reconditioned Hearing Aids—Hear Now is a nonprofit organization that collects used behind-the-ear hearing aids and distributes them throughout the United States to hard-of-hearing people with limited financial resources. The reconditioned hearing aids are fitted and dispensed by a nationwide network of health care professionals who waive their usual fitting fees for qualified applicants.

Hear Now has an extensive collection of used hearing aids. It is developing a national registry of community-based hearing aid assistance programs. Whenever possible and appropriate, Hear Now will refer individuals to local programs in their own communities. (See ***Helpful Resources.***)

HELPFUL RESOURCES

To donate a used hearing aid or to obtain more information about the nationwide program that distributes reconditioned behind-the-ear hearing aids, contact **Hear Now** at 9745 E. Hampden Ave., Suite 300, Denver, CO 80231-4923; (800) 648-HEAR or (303) 695-7797 (voice/TTY).

HOUSES

Although they may not realize it, homeowners who protect their houses with regular applications of preventive maintenance are also protecting the planet that all of us call home.

A house represents an immense amount of natural resources and energy. As an example, a "typical" three-bedroom house contains 10 to 15 trees worth of lumber, nearly 5,000 square feet of drywall, and more than 40 cubic yards of concrete; also good amounts of copper and other metals found in plumbing, hardware, nails, and appliances; not to mention ample quantities of oil-derived floor coverings, plastics, paints, varnishes, etc.

In addition to these materials, a house embodies a staggering amount of energy, including the energy that was needed to extract the natural resources from the earth; the energy required to transport raw materials to the factory or mill where still more energy is used to process them into finished materials and building parts; the energy required to transport finished materials to the building site; and finally, the energy used to construct the house.

Home Maintenance Pays Big Environmental Dividends—By preventing expensive repairs, regular maintenance conserves building materials that would otherwise be needed for rehabilitation. When you conserve building materials, you also diminish environmental degradation caused by mining and lumbering operations, save gobs of energy (including your own), and help minimize the trail of air pollutants left behind when fossil fuels are burned. This means less smog and less carbon dioxide (the primary greenhouse gas). In the case of coal-derived energy, it also means less acid rain and less strip mining. You get the picture. In the long run, caring for your home contributes to cleaner air, healthier forests, crops, lakes, wildlife, and people.

Looking for Trouble—A house contains a complex of systems (heating, electrical, plumbing, sewer, etc.), which should be inspected annually. Following a maintenance schedule makes it possible to avoid some, though probably not all, rude surprises. It is better to have your furnace inspected and serviced during the summer months, for instance, than to have it break down on you in the depths of winter. Similarly, by checking your roof at least once a year, you'll know when it may be time to install a new one. Paying several thousand dollars for a new roof may seem expensive, but it's cheap compared to the cost of repairing hidden water damage to your home's wood framing caused by a leaky roof or clogged gutters and downspouts.

(Also see: ***Furnace Filters; Refrigerators; Building Materials; Lumber.***)

HELPFUL RESOURCES

These books provide step-by-step instructions for annual house inspections that can help you find problems early, perform basic upkeep, even implement minor repairs:

Year-Round House Care: A Seasonal Checklist for Basic Home Maintenance by Graham Blackburn (Consumer Reports, 1991, $14.95).

New York Times Season-by-Season Guide to Home Maintenance by John Warde (Times Books/Random House, 1992, $25).

KITCHEN TOWELS

The Siren Call of Convenience—Go ahead. Reach up and rip off a paper towel. Dry your hands. Mop up that spilled juice. Then toss the soggy spent wad of paper in the trash.

Using paper towels is a habit. And habits are powerful forces. When you multiply your own household's daily paper towel habit by 365 days a year and millions of consumers, it really adds up. It has been estimated that Americans use over 27 million trees worth of paper towels each year. Towel by towel, we are pulping our forests into throwaway products.

It would be nice if paper towels could be recycled. Unfortunately, even though some paper towels contain recycled fibers, the wet-strength additive in paper towels renders them unrecyclable.

A Change of Habit—Go ahead. Break the paper towel habit by splurging on a rainbow array of colorful cotton kitchen towels. If you buy them by the dozen, cotton towels can cost as little as $2 a piece, and each towel could easily outlast $30 worth of paper disposables.

To ensure that you always have a clean towel for drying freshly washed hands, launder kitchen towels regularly. When you toss them in with the rest of your launderables, you add virtually nothing to daily chore time, and the expense is negligible. You may want to designate one towel specifically for drying clean hands, another for wiping countertops, etc. Use a sponge to wipe up spills.

Pass these earth-wise habits on to your kids. With a touch of positive reinforcement, even stubborn spouses can acquire the cloth towel habit. (My husband now says he actually prefers cloth towels to paper.) Of course, one needn't go cold turkey on paper towels. There's nothing wrong with keeping a roll on hand. Just save it for special occasions.

LUMBER

Looking beyond the Trees—The simple dollar value of harvestable timber is only the most obvious of many economic and ecological blessings that forests bring. By absorbing carbon dioxide, woodlands play a vital role in stabilizing climate and slowing global warming. By absorbing rainwater and gradually releasing it into streams, forests prevent flooding and soil erosion. Forests also protect fisheries in rivers and lakes, harbor millions of plant and animal species, and provide recreation areas for humans.

Forests for the Future—According to the Worldwatch Institute, roughly one-fifth of Earth's forests have vanished in the last 45 years—felled for timber, paper products, and fuel or to make way for cropland, cities and suburbs.

So that future generations may enjoy the fruits of the forests as we do, it is essential that we develop sustainable forestry practices. Modifying harvesting techniques can certainly help, but it is unlikely to be enough. We'll also need to moderate our voracious appetite for wood products.

Wood Is Too Valuable to Waste—Every year, America discards an estimated 12 million tons of wood—enough, according to the Center for Neighborhood Technology, to frame a house for each person in the country.

For some entrepreneurs, all this waste is a potential gold mine. In cities across the nation, discarded shipping pallets that once went to landfill are now being reclaimed by pallet recyclers who repair and resell them. The result? A booming new industry that sells an estimated $3 billion of remanufactured pallets annually.

Instead of smashing old buildings to splinters and hauling the remains to landfill, some salvage companies have been developing gentler "deconstruction" techniques that involve methodically picking structures apart to rescue as much lumber and re-usable building parts as possible.

At the Presidio, a former military base in San Francisco which is being converted to a national park, officials plan to auction unusable buildings to demolition contractors for the value of the salvageable materials they contain.

Lumber and building parts salvaged from deconstruction projects like the Presidio often end up in used-lumber yards or used building materials stores, where contractors and the general public can purchase them for a fraction of the price of new materials.

Salvaged Wood Is Beautiful—While bargain hunters may be enticed by the low prices of salvaged wood, some builders praise old wood for its stability. (Having been seasoned for several decades in its first building, salvaged wood is usually denser and dryer than freshly cut lumber.) Cabinet and furniture makers cherish the magnificent grain structure found only in the super-large old growth timber that has become so scarce. And architects prize the "ambience and character" that previously used lumber can bring to a project. Microsoft billionaire Bill Gates's giant new luxury home in the Pacific Northwest incorporated six barns worth of weathered wood from dilapidated old farm buildings.

Caveat Emptor—Used lumber dimensions are not always uniform, so builders should plan accordingly. (Planing the wood to size is a hazardous task best left to the experts.) The majority of used lumber has not been graded, so it cannot be used for projects that must be permitted and inspected.

Inspect used lumber carefully. Avoid wood that smells of chemicals or that is coated with lead-containing paint.

Safety First—Because high temperatures and humidity or heavy loads can impair wood strength over time, some experts recommend re-using wood only for nonstructural purposes such as fences, decks not designed to carry a heavy load, siding, paneling, cabinets, furniture, pallets, and shipping crates.

If you plan to saw the lumber yourself, use protective eyewear and watch out for old nails and metal fasteners. Even if it has been de-nailed, used lumber can still have metal in it (sometimes broken off and hidden below the surface). (Also see: ***Building Materials.***)

HELPFUL RESOURCES

1. Outlets for used lumber are gradually increasing in number. Look for them in the Yellow Pages under such headings as "Used Lumber," "Salvage," "Architectural Salvage," or "Demolition Contractors."

   For the name of a company that sells used lumber in your area of the U.S. or Canada, contact the **Used Building Materials Association** at (204) 489-2739. (See Appendix B, ***Organizations.***)
2. ***Cut Waste Not Trees: How to Use Less Wood, Cut Pollution and Create Jobs*** explores a multitude of creative and profitable ways to reduce wood consumption and preserve forests. It is available for $7.50 from the publisher, **Rainforest Action Network**, 450 Sansome St., Suite 700, San Francisco, CA 94111; (415) 398-4404.

LUNCH TOTES

The students at Mendon Center Elementary School in Pittsford, New York, learn early to make re-use a way of life. At the beginning of the Fall semester, parents receive a note outlining environmentally sound methods for packing "no garbage" lunches. The guidelines encourage parents to purchase re-usable lunch boxes and bags; to utilize re-usable plastic containers instead of single-use plastic bags, food wrap and foil; and to use a washable insulated beverage container in lieu of disposable juice boxes, bottles, and cans.

Packing a homemade lunch isn't just for kids. It's for everybody who wants to eat well or save some money for a rainy day. Preparing your own lunch gives you greater control over what you eat, and that can mean a tasty, nourishing midday meal with significantly less cholesterol and sodium than standard restaurant fare. Homemade lunches make for a healthier planet too. Just think of all the natural resources and energy you'll conserve when you're no longer tossing single-use take-out containers in the office trash bin.

Lunch boxes have come a long way since the 1950's, when I carried a metal box adorned with likenesses of Roy Rogers and Dale Evans. The lunch carriers of the nineties are made of durable plastic or washable nylon canvas, and they come equipped with insulation that keeps food warm—or cold—far better than the metal boxes of my youth. You won't find old-fashioned glass vacuum **Thermos®** bottles inside the modern lunch tote. Those fragile vessels have largely been replaced by more durable beverage containers lined with unbreakable plastic or stainless steel.

Today's lightweight lunch totes come in a variety of shapes and sizes: standard sized bags and boxes for individuals; larger capacity carriers and backpacks for big eaters or picnickers. For those who would rather not carry a separate lunch tote, there are compact compartmented food carriers not much bigger than a large paperback book that fit neatly inside a purse or backpack.

HELPFUL RESOURCES

Look for sturdy lunch totes and re-usable plastic food and beverage containers at supermarkets, large drugstores, super stores and hardware stores.

MAGAZINES

It has been estimated that Americans discard approximately two million tons of magazines every year. Fortunately, we needn't discard magazines any longer. We can recycle them or, better yet, with a little ingenuity we can make them available to others who might enjoy reading them—and *then* recycle them. So before you toss your magazines in the trash or recycle bin, consider offering them to someone who might want to re-use them.

Read, Reread, and Re-reread—People with time on their hands always seem to enjoy reading material, current or not. Many hospitals and convalescent homes welcome a variety of general interest magazines. Or perhaps a medical or dental office could use them in the waiting room.

At my local recycling center, a great many of the magazines dropped off for recycling are retrieved and re-used by adventurous patrons who climb into the magazine dumpster to rummage for reading matter. A neighbor of mine brings her teenagers to the magazine dumpster to dig for buried issues of *Seventeen* and *Sports Illustrated.* Other dumpster divers take home issues of *The New Yorker* and *National Geographic,* computer mags, fashion rags, even *Playboy.* (The dumpster is uncensored.)

If dumpster diving is not your cup of tea, plenty of public libraries maintain an informal magazine swap shelf, where patrons can leave their old magazines for others to take home.

Some used-book stores buy and sell used magazines. Half-Price Books, a chain of used-book stores with 55 stores in 8 states, buys and resells current issues of many general interest journals. They also stock back issues of *National Geographic* and welcome virtually any items published before 1940. Used magazines that don't sell eventually get donated to local organizations. In Berkeley, California, for example, Half-Price Books donates unsold items to the county prison libraries.

Day care and elementary school teachers often utilize magazines for classroom crafts. Children can cut pictures from magazines to create collage art or to make greeting cards for Christmas, birthdays, Mother's Day, Valentine's Day, Easter, etc.

Colorful pages from gardening magazines are handy for wrapping tiny gifts.

If several people in your office receive the same journal, you can cut waste while saving dollars and resources by ordering a single subscription, then using a routing slip to circulate the magazine among multiple readers.

MAPS

The world is constantly changing. Every year, new roads are built, new suburbs spring up, governments topple, and national boundaries are redefined. And to reflect these emerging realities, new maps must be drawn.

Meanwhile, what happens to the old maps that are now obsolete? Well, that's up to you. If they're already tattered and the fold lines have begun to perforate, you may simply want to recycle them as "mixed paper" (provided mixed paper recycling is available in your community). But if they're in decent condition, even old maps have their uses.

Outdated maps can be re-used to wallpaper a child's room, a den, or a hallway. This is a perfectly legitimate, if temporary, way to cover cracked plaster you haven't yet gotten around to patching.

Maps—even outdated ones—are great educational tools. And plenty of teachers would love to receive your old maps, especially ones from *National Geographic* magazine. Outdated maps are a great way to demonstrate to students how national boundaries change over time. Ask schools and teachers in your area if they would like your old maps.

Some thrift shops welcome donations of *National Geographic* maps, and a few libraries also accept such maps for fund-raising sales. Please call first, though, to make sure your maps are a "wanted" item.

MATTRESSES

Make Your Mattress Last—If you want your mattress to enjoy a long and useful life, don't let your kids use it as a trampoline and be sure to follow the manufacturer's maintenance instructions. Most manufacturers recommend rotating a mattress periodically by turning it over and reversing it end-to-end. This maneuver helps minimize the sagging body impression that develops over time.

When you buy a new mattress, the delivery charge often provides for free hauling away of the old one. After it leaves your house, the discarded mattress usually goes to landfill, where—because the springs do not compact readily—it takes up a lot of precious space.

Recycling Is Great—Sometimes to conserve landfill space, mattresses destined for the dump are first stripped of readily recycled components such as steel frames and springs. One unusual mattress recycler in Dublin, California, goes a step further and recycles nearly all the constituent materials. The urethane foam padding is sold to firms that shred it and make it into rebond, a carpet underpad. The frame wood is sold for kindling, and plywood from box springs becomes boards for children's bunk beds.

But Re-use Is Better—If you are buying a new mattress and the old one is still sleepworthy, consider offering it to a charitable thrift shop such as the Salvation Army. This will keep it out of landfill and provide someone with decent bedding for several years. Since not all charity thrift shops accept used mattresses, you may need to make a few phone calls to find one that will come out and pick it up.

If your old mattress has sound springs and a cover that is neither torn nor stained, the charity can sanitize and resell it "as is" in its thrift store. (Most states, though not all, require used mattresses to be sanitized before they are resold to the public.)

If the springs are sound but the cover is stained, the charity may turn your old mattress over to a renovating company, where the old cover (and possibly the padding as well) will be replaced and the whole mattress, if required by state law, will be sanitized. Renovated mattresses are typically sold at discount and bargain outlets for significantly less than "brand new" ones.

NEWSPAPERS

I used to think my husband was cheap because whenever we went out for breakfast on Sunday, rather than buy a paper from the newsstand he would read the papers left behind by previous patrons. But then I read that supplying Americans with their Sunday editions gobbles up an entire forest—over 500,000 trees—each week. Since then, I've revised my judgment. The spouse is no longer cheap. He's environmentally considerate.

Sometimes Recycling Is Better Than Re-using—For years, resourceful householders have improvised creative ways to re-use yesterday's news. They've wadded up old newsprint and used it for cleaning windows, spread it on the kitchen floor to house train puppies, ripped it into strips for cat litter box fill, and rolled it into fuel logs for the fireplace.

But let's face it, the daily news is essentially a single-use throwaway product. And most re-uses are little more than a brief detour—albeit a useful one—on the way to the trash heap. In contrast, recycling old newspapers yields feedstock for the creation of "new" newsprint, moderates the demand for virgin pulp, and allows whole forests to stand. In 1994, Americans saved roughly 85 million trees from the chain saw by recycling their newspapers.

NURSERY EQUIPMENT and FURNITURE

Cribs, strollers, high chairs, changing tables, toy chests, car seats, play pens. Quite often you don't even have to go looking for these items; they come to you. When friends and relatives learn that you are expecting, they offer you the stuff their kids have outgrown.

If the nursery items don't come to you, you can go out and find them at resale boutiques for children, at thrift stores, yard and garage sales, or in the classified ads of the local newspaper. Keep your eyes open, and you'll notice ads for baby equipment posted at the grocery store, the public library, or alongside the office water cooler.

Durably built nursery equipment can serve a succession of children—with one caveat: some hand-me-downs and older equipment may have unsafe features that have been designed out of more recently manufactured products. Take cribs, for example. Prior to 1973, when the federal government issued stringent safety standards, faulty crib designs were responsible for thousands of infant injuries (even deaths) every year.

With a little knowledge, however, you can steer clear of potential perils and select safe equipment for your child. The information you'll need is available for free in a booklet published by the U.S. Consumer Product Safety Commission (CPSC). (See ***Helpful Resources.***)

Also see: ***Child Safety Seats.***

HELPFUL RESOURCES

1. Written for parents who already have hand-me-down equipment or are shopping for used nursery furnishings, ***The Safe Nursery: A Booklet to Help Avoid Injuries from Nursery Furniture and Equipment*** identifies unsafe features that parents should avoid when shopping for cribs, playpens, high chairs, gates and enclosures, toy chests, walkers, changing tables, and other nursery items.

The booklet's equipment safety checklist can also be used to inspect nursery equipment now in use in your home and to keep it in good repair.

For your free copy, call the CPSC toll-free hotline: (800) 638-2772. Or write to **Publication Request, Office of Information and Public Affairs,** U.S. Consumer Product Safety Commission, Washington, DC 20207.

2. Even brand new equipment sometimes proves unsafe. If enough consumers report the hazard to the CPSC, it may be recalled. You can obtain a free list of current "Recalls and Corrective Actions for Toys and Children's Products" from the Consumer Product Safety Commission at the phone and address listed in the paragraph directly above.
3. Check out "How to Find a Good Safe Crib" in the May 1993 issue of ***Consumer Reports*** magazine. This article presents guidelines to help parents recognize and avoid unsafe cribs. Look for back issues of *Consumer Reports* in the reference section of your local public library.

PACKING PELLETS

Made of foamed polystyrene and often formed in the shape of peanuts or shells, packing pellets are widely used to cushion and protect products during shipping. Unfortunately, manufacturing these handy little pellets generates a hefty volume of hazardous chemical wastes. In 1986, when the EPA ranked the twenty chemicals whose production generated the most hazardous waste, polystyrene was fifth on the list.

Manufacturing polystyrene foam also generates air pollution. The culprits here are the blowing agents used in the foaming process. Because they cannot be readily captured and recycled, blowing agents are usually discharged into the air. Pentane, one of the most widely used blowing agents today, contributes generously to smog and global warming.

Recycling Is Good—If uncontaminated, polystyrene can be recycled. Although recycling conserves oil that would otherwise be needed to produce virgin polystyrene, it does not reduce air-polluting emissions. In fact, it has been estimated that recycling post-consumer polystyrene actually produces about 8 percent *more* air polluting emissions than manufacturing with virgin resins.

Re-use Is Better—Foam packing pellets have been dubbed villains in landfills because they do not biodegrade. It is this very durability, however, that makes it possible to re-use them over and over again.

The beauty of re-using these sturdy pellets is that it reduces the demand for new ones, thus minimizing hazardous chemical wastes and air pollution generated by the manufacturing process. By decreasing the demand for new polystyrene, which is petroleum-derived, re-using foam packing pellets conserves oil and helps diminish environmental damage caused by oil and gas drilling.

Opportunities for Re-use—Individual householders can either re-use pellets for their own packages or turn them over to packaging services and postal service stores, which are usually very glad to re-use clean, dry foam pellets. (See ***Helpful Resources.***)

HELPFUL RESOURCES

Packing pellets can be returned for re-use to over 3,200 collection sites across the United States. For the location nearest you, contact the **Plastic Loose Fill Producers Council** at (800) 828-2214. Or call a local outlet for one of these postal services or packaging stores:

Associated Mail and Parcel Centers
Mail Boxes, Etc.
Packaging Store/Handle with Care
Pak-Mail

PAINT

Most of us stash leftover paint in the basement or garage because, well, you never know when it might come in handy. But paint does not keep forever. And that opportune use might not arise until long after the leftovers have dried, soured, or frozen into unspreadable sludge. Wouldn't it be better to find a use—or someone who can use it—while the paint is still fresh and usable?

Find a Use and Paint It—The easiest way to dispose of extra paint is to use it up. Provided the colors are compatible, you can combine various leftovers and use the mixture to paint just about anything. When I combined three partially empty cans of various off-white interior latex paints, the result looked just as sharp as if I'd gone to the paint store and paid $25 for a fresh can.

Don't Mix Oil and Water—It's perfectly legitimate to mix any latex paint with any other latex paint; or one oil-based paint with another oil-based paint. But don't mix oil-based paints with latex (water-based) paints—the old adage that "oil and water don't mix" is still as true as ever.

For outdoor projects, it's OK to combine interior and exterior grade paints, although such combinations may not be quite as durable as mixtures containing only exterior grade paints.

For indoor painting, it's best to stick with interior grade paints, as exterior grades may contain mercury or biocides, which can pose a health risk when used indoors. Any paint with a label that indicates it kills mildew or fungus is also likely to contain mercury. Additionally, there's mercury in about 30 percent of interior latex paints produced before August, 1990, when the U.S. Environmental Protection Agency banned its use in interior latex paints. (See ***Helpful Resources***.)

Read Labels Carefully—In 1978, the U.S. Consumer Product Safety Commission virtually banned lead in paints. If you have reason to believe your paint was produced before 1978, don't use it. Take it to your community's next hazardous household waste collection event. (See ***Helpful Resources***.)

Finding Users—If you can't use it yourself, offer your leftover paint to friends and neighbors, to churches, theater groups, schools, civic organizations, home rehabilitation programs like Christmas in April, or graffiti abatement programs.

Another way to keep usable paint from going to waste is to drop it off at a paint exchange, where other users in need of paint can cart it home for free. In Petaluma, California, a non-profit organization known as Garbage Reincarnation operates a paint exchange that gives away about 400 gallons of paint each month. The rules of the exchange are simple: anyone can drop paint off, and anyone can take paint home. All paints must be in their original container, with the label intact. Each can must be at least one-third full.

The Central Vermont Planning District periodically sponsors a one-day paint Drop & Swap, where citizens can bring unused paints, stains, paint strippers, thinner, and brush cleaners. Paint that is deemed hazardous is earmarked for proper disposal. The rest is displayed on a re-use table, and everyone is invited to take what they can use. A pilot Drop & Swap event in 1989 diverted nearly 800 gallons of old paint to new users.

Recycling Is Still an Option—Of course, you can always deliver your unwanted latex paints to a paint recycling program, which will filter, blend, and repackage the paint in a new container. Recycling is less resource efficient than exchanging it directly with a new user, but it's certainly better than throwing it away or letting it spoil in the garage.

Don't Buy a Gallon When a Quart Will Do—To ensure that you buy only the amount of paint you need, measure the area you plan to paint (with accurate deductions for windows and doors), and ask the retailer to help you calculate how much you'll need.

If you have only a small amount left over, you can save it for touch-ups later on. Be sure to label the container clearly and store it out of children's reach.

HELPFUL RESOURCES

1. Want to organize a paint swap for your community? The Vermont Agency of Natural Resources publishes ***Guidelines for Conducting Paint Drop & Swap Events.*** It is available for $5 from **Vermont Solid Waste Division,** 103 S. Main St., West Building, Waterbury, VT 05676.
2. Wondering if there's mercury in that interior latex paint you bought a few years back? Information about the mercury content of latex paints manufactured since 1988 may be obtained from the EPA's toll-free **National Pesticide Hotline:** (800) 858-7378.
3. Many communities sponsor collection programs for hazardous household wastes. Such programs are a proactive attempt to keep toxic materials out of municipal landfills, where they may eventually leach into the surrounding groundwater and pollute local drinking water supplies. Collected hazardous wastes are transported to facilities specifically designed to manage toxic materials as safely as possible.

   To find out if your community has a hazardous household waste collection program, call a local recycling agency, the city sanitation department, or the health department. In many areas, such collections may be managed by the county sanitation or health department.

PANTYHOSE

Supple, durable, and surprisingly strong, old pantyhose are more re-usable than you might think.

Bundle Newspapers for Recycling—Cut the panties off the hose; cut each leg in two lengthwise strips; tie the strips together end-to-end. This creates a knotted rope of pantyhose, which can be used as string for bundling newspapers.

Strain Lumps from Paint—Stretch a single thickness of pantyhose over a clean container, secure it with a sturdy rubber band, and pour the lumpy paint in *very slowly.* You can use either the panty or the leg, but for wide-mouth containers the panty portion works best.

In the Garden—Cut strips of pantyhose and use for tying garden plants to stakes and fences. Because they're made to stretch and flex, the hose won't cut into the delicate stalks of plants as they sway in the wind or grow toward the sun.

Create Stuffed Animals—Some senior citizen groups or church groups utilize retired nylons to make stuffed animals or other crafts. Call a few churches or senior groups to find out who might want your discards (after you've laundered them, of course).

PAPER

Congratulations, Recyclers!—Your efforts are making a difference. In 1994, Americans recovered record amounts of paper and paperboard products—nearly 29 million tons of the stuff, or about 500 million trees worth.

But Recycling Is Not Enough—World production of paper products is projected to double by the year 2010. To meet this burgeoning demand, forest economists anticipate a future of steadily expanding timber harvests. At best, even higher levels of recycling will merely slow the rate at which tree harvesting continues to increase.

Sheet by Sheet, Tree by Tree—At the touch of a button we can fax, photocopy, or print—one page or a thousand pages. It's so easy and automatic that we don't give it a thought. Yet every single sheet of paper we use carries an environmental price tag.

For starters, there is the loss of valuable ecological services that forests perform. When more than a billion trees are felled annually to feed our nation's voracious appetite for paper products, the deforested areas are no longer able to absorb carbon dioxide, prevent flooding and soil erosion, or harbor millions of plant and animal species.

Transforming all those trees into paper is a far more energy intensive process than most of us realize. It has been estimated that it takes about 10 kilowatt-hours to manufacture a ream (500 sheets) of paper. As one of the most energy-consuming industries in the country, papermaking is a major producer of sulfur dioxide (the chief cause of acid rain) and carbon dioxide (the primary greenhouse gas).

From Chlorine Bleach to Dioxin—When totting up the environmental costs of papermaking, it's important to understand that turning brown-colored wood pulp into bright white paper calls for a lot of bleaching. Although a few mills rely on hydrogen peroxide for whitening, most U.S. mills utilize chlorine, which reacts with compounds in the wood to produce highly toxic synthetic chemicals.

Every year, the U.S. pulp and paper industry discharges millions of pounds of toxic chlorine compounds into rivers and coastal waters. Included in this toxic soup is dioxin, a notorious family of carcinogens now believed to suppress immune systems and impair reproductive ability in humans and animals.

Perhaps the scariest thing about dioxin is its resistance to biological breakdown. Don't expect this stuff to biodegrade or vanish anytime soon. In fact, dioxin is bioaccumulative. That is, it's passed up the food chain and stored in the fat cells of each organism that ingests it. In this way, dioxin absorbed by a fish that swam in a polluted lake is deposited in the fatty tissues of a breast-fed baby whose mother unwittingly ate the contaminated fish.

Every Sheet of Paper Has Two Sides. Use Both of Them—One of the most effective paper reduction measures available to just about everybody is re-use. By using both sides of each sheet we can minimize the environmental impact of our paper-intensive life style.

1. Use the clean side of discarded papers to draft memos and reports, take telephone messages, make notes, or for children's homework.
2. Turn print overruns, computer printouts, outdated forms and stationery into note pads or notebooks.
3. Print and copy all multi-page documents on both sides. Double-siding cuts paper costs in half. It has been estimated that double-sided copying can save small companies or departments about $700 annually in paper costs.

   Bank of America saved more than $1 million with paper reduction measures that included printing customer statements on both sides of each page.

A software development company in Mountain View, California, stocks several copiers with previously used one-sided paper. Employee announcements, file copies, and other internal documents are printed on these sheets.

Make Less Paper Count for More

4. Paper that has been used on both sides is ready for recycling. It can also be shredded and utilized as packing material. (Small shredders may be purchased at office supply stores for about $50.)
5. Although laser printer manufacturers warn against loading the paper tray with previously used paper because it tends to cause paper jams, quite a few folks do it anyway and claim it works fine as long as the paper isn't crinkled or dog-eared. I utilize once-used paper in my inkjet printer regularly. (Yes, every now and again it *does* jam, but such jams are easily cleared.)
6. When faxing documents, use cover sheets that have been pre-printed on the clean side of previously used paper.
7. Instead of sending individual memos to each employee, route one hard copy to several readers or utilize electronic mail. Announce new policies in department meetings and post individual copies on a centrally located memo board in each department. ***Tip:*** Keep posted memos short. Large print will make them more readable.

PERSONAL COMPUTERS

The Dark Side of Personal Computing—Personal computers have revolutionized the way we perform a multitude of tasks—from typing term papers to communicating with people on the other side of the globe. Alas, these marvelous devices also have a dark side, which we ignore at our peril. Using PCs consumes horrendous quantities of paper, ink, plastic cartridges, and electricity. And manufacturing PCs generates dismaying amounts of highly toxic pollution.

Making circuit boards and microchips, disk drives and other PC components involves a veritable alphabet soup of toxic chemicals. In Northern California's Silicon Valley, the center of the world electronics industry, manufacturers produce over 100,000 tons of hazardous waste annually. Unfortunately, much of this waste has not been properly handled, and as a result Silicon Valley has become one of the most polluted places on Earth, with twenty-nine Superfund sites.

Such pollution problems are exacerbated by the relentless advance of computing technology, which renders many PCs obsolete long before they actually wear out. According to a recent Carnegie-Mellon University study, every year U.S. businesses jettison an estimated 10 million PCs for faster and fancier ones. This chronic obsolescence keeps things lively on the assembly line and probably creates quite a few jobs; it also begets a good bit of pollution.

On the Bright Side—Recognizing that many castoff PCs still have plenty of computing life left in them, refurbishers purchase thousands every year from upgrading corporations. They test and clean the PCs, then sell them to schools, government agencies, exporters, and thousands of U.S. retailers nationwide. By moderating the demand for brand new equipment, those who purchase refurbished PCs are helping to minimize manufacturing-caused pollution.

Who Buys Used PCs?—Individuals and small businesses that need a PC primarily for word processing or basic spreadsheets often find that last year's technology provides all the computer power they want or need. These users don't require all the latest bells and whistles, and they'd rather not pay for them either.

Savvy shoppers who opt for a PC that's already been thoroughly road-tested can get a lot of computing power for their buck. According to one survey, 62 percent of the 2.4 million used PCs snatched up by American consumers in 1995 sold for less than $500.

Figuring out what is a fair price to pay for a used PC can be baffling. Once you know the model and make you want to buy, you'll need to phone several used computer stores, then compare those prices to listings for similar equipment at a computer exchange. (A computer exchange arranges sales between sellers and buyers of used equipment.) (See ***Helpful Resources.***)

Replacement Parts—Before purchasing a used PC, make sure that replacement parts are available, in case your machine ever needs repairing. Many refurbishers maintain a ready supply of used parts, and parts can be purchased from used parts suppliers and brokers. Repair shops should be familiar with the used parts marketplace.

Warranties—Worried about getting stuck with a lemon? Choose a reputable dealer and insist on a good warranty that covers both parts and labor. Many stores offer only a 30-day warranty, but some provide 90-day warranties.

Upgrading—If you already own a PC but need additional computing power, you may want to consider upgrading rather than buying a new one. Some manufacturers offer PCs that can be easily upgraded by replacing the old central processing unit (CPU), also known as the "processor," with a new one. But even if your PC wasn't designed to accommodate a new CPU, you can often boost performance by upgrading other hardware components. There is a good selection of books on upgrading PCs, and one of these books can help you determine whether a hardware upgrade makes sense for your machine. (See ***Helpful Resources.***) Computer repair shops are another good source of advice.

Plan Ahead—If you need to upgrade several components, it may be more cost-effective to purchase a new PC (or a used one that has the capabilities you need). So that a new computer can serve you as long as possible, make sure it can be readily upgraded to the next-level CPU, can accept up to 32 megabytes of memory, and has at least one expansion slot to accommodate an additional adapter board.

Avoiding the Trash Heap—When buying a new system leaves you with an older one that still works, don't let it go to waste. There are plenty of people who could put it to good use.

Offer it to a friend or relative. Or try selling it through a newspaper ad or computer bulletin listing. It may be more convenient, however, to market it through a computer exchange. (For more information about computer exchanges and how they work, see ***Helpful Resources.***)

Alternatively, you may collect a nice tax deduction by donating your machine to a local school or nonprofit organization. (See Appendix A, ***Donate Instead of Dump.***) Many groups would be overjoyed to take a functional PC off your hands. A few even accept nonworking PCs and refurbish them. Some of these groups are listed under ***Helpful Resources.***

If it's no longer functional, offer your machine to a computer maintenance and repair business. They may be able to dismantle it and re-use individual parts to fix similar machines.

A Clean PC Lasts Longer, Works Better—Believe it or not, simply keeping your computer clean can prevent a host of unwelcome repairs and help it enjoy a long and useful life.

Dust is the kiss of death to computers. When it gets inside the case (sucked in by the cooling fan), dust causes metal parts to overheat and eventually to fail, increases energy consumption, and shortens the life of the system. Whenever you're not using it, keep a dust cover on your PC. You can purchase plastic covers at a computer store, but an old pillow case will do just fine. If a peek under the hood reveals dust laden innards, any PC service center can clean it for about $25.

HELPFUL RESOURCES

1. Want to know how to keep your PC up and running for as long as possible? ***The Green PC*** by Steven Anzovin (1993, Tab Books, $9.95) is full of earth-friendly ways to keep your PC in top condition and boost its performance with simple hardware upgrades.
2. Want to learn more about buying used PCs? Alex Randall's ***Used Computer Handbook*** (1990, Microsoft Press, $14.95) provides sound shopping advice and simple procedures for test-driving a used machine before purchasing.
3. Wondering if your PC can be upgraded? These books can help:

   PC Magazine's Guide to Upgrading PC's by Dale Lewallen (Ziff-Davis, 1996, $34.99).

   The Complete PC Upgrade & Maintenance Guide by Mark Minasi (Sybex, 1996, $54.95).
4. For a fee, a computer exchange will arrange sales between buyers and sellers of used PC equipment. Individuals with equipment to sell can list it on an exchange's database. Potential buyers can request prices on available equipment. Listed prices are "asking" prices; actual sales prices, after bargaining, may be lower. In addition to the exchanges listed below, you may also find local exchanges in the Yellow Pages.

   United Computer Exchange: Powers Ferry Rd., Suite 307, Atlanta, GA 30339; (800) 755-3033; 2110.

   American Computer Exchange: 190 Sandy Springs Place, Suite 102, Atlanta, 30328; (800) 786-0717.

   National Computer Exchange (NaComEx): 230 Park Ave., Suite 1000, New York, NY 10169; (800) 622-6639.

5. These charitable organizations accept functional PCs and route them to schools or nonprofit groups:

   National Christina Foundation collects all models and distributes them to organizations serving the disabled and the disadvantaged. 591 W. Putnam Ave., Greenwich, CT 06830; (800) CHRISTINA or (203) 622-6000.

   The Computer Recycling Center, Inc. accepts equipment in any condition (working or nonworking), refurbishes and distributes it to California schools; also sells low cost PCs to students, low income individuals and families every Saturday morning. 1245 Terra Bella Ave., Mountain View, CA 94043; (415) 428-3700.

   Non-Profit Computing, Inc. analyzes the needs of nonprofit organizations and government agencies (including public schools), then carefully matches with appropriate equipment. This group also offers workshops, pro bono consulting services, clinics, and training classes. For information on how to qualify to receive donated equipment or software, send a stamped self-addressed envelope to 40 Wall St., Suite 2124, New York, NY 10005-1301. To donate hardware or software, call (212) 759-2368.

   East West Education Development Foundation places donated equipment in schools in the former Soviet Union and Eastern Europe. 23 Drydock Ave., 3rd floor, Boston, MA 02210; (617) 261-6699.

   Computers and You uses donated PCs to teach computer skills to the homeless and low income community in San Francisco's Tenderloin district. 330 Ellis St., San Francisco, CA 94102; (415) 922-7593.

   Computers for Schools coordinates the distribution of donated used computers to California schools. Contact executive director Diane Detwiler at 470 Nautilus St., Suite 300, La Jolla, CA 92037; (800) 939-6000 or (619) 456-9045.

   National Campaign for Education promotes education and computer literacy by distributing donated PCs to qualifying nonprofit organizations. 230 Park Ave., Suite 1000, New York, NY 10169; (800) 622-6639.

PERSONAL COMPUTER SUPPLIES

Using personal computers consumes more resources than most of us realize. Each year, PC users devour more than 9 million trees worth of paper, a billion floppy disks, 60 million laser toner cartridges, 100 million ribbons, and heaven-knows-how-many inkjet cartridges.

Is there anything PC users can do to curb these earth-battering excesses? You bet there is.

Laser Toner Cartridges—Sending empty laser toner cartridges to a recharger for refilling benefits the planet in a big way. Not only does it keep a 4-pound hunk of plastic resins, carbon, selenium, and other metals from hogging landfill space, it conserves much of the energy and labor originally used to manufacture the cartridge.

Cartridge recharging makes economic sense too. A remanufactured cartridge costs about $50, saving you $25 to $40 over the price of a brand new one.

A reputable recharger does quite a bit more than just refill an empty cartridge. Each cartridge is disassembled and inspected for signs of wear. Worn out parts are replaced with new or reclaimed ones. The cartridge is then refilled, reassembled, and tested.

For companies that use a lot of cartridges, local rechargers will often pick up empties and deliver refilled units, just like the milkmen of bygone days. But individuals and low volume businesses generally have to provide their own transportation. If this is not convenient, anyone can purchase recharged cartridges off the shelf at stationery stores, large office supply outlets, computer stores, or through the mail from computer supply catalogs.

Since there are now more than 8,000 recharging companies in the U.S., it shouldn't be hard to find one nearby. (Look in the Yellow Pages under "Computer Supplies.") Ask a recharger how long they've been in business and request local references. If others are well satisfied with the company's work, there is a good chance you will be too.

Although quality standards in the recharging industry have

increased dramatically in recent years, it's always prudent to inquire about the recharger's warranty. Rechargers who follow the industry standard will replace any cartridge that fails.

Even if you rely exclusively on brand new cartridges, there is no excuse for tossing empties in the trash. Most of the major laser printer manufacturers will take back spent cartridges for recycling. Some even pay postage. Look for mail-back instructions inside the carton. Although they don't recharge and refurbish the whole cartridge, some manufacturers *do* salvage a few parts and re-use them in "new" cartridges.

Better yet, many local rechargers will pay cash for your empties (from $1 to $7 each, depending on the make and model). By installing a long-life photoconductive drum, they can recharge them up to six times.

Inkjet Cartridges may be refilled 2 to 5 times. After that, the cartridge's printhead surface begins to clog up and print quality deteriorates.

There are several ways to get inkjet cartridges refilled. Some of the same companies that recharge laser toner cartridges also refill inkjet cartridges. Before adding a fresh supply of ink, a reputable refiller will clean each cartridge and inspect it for cracks and signs of wear. The charge for refilling is typically about 30 to 50 percent less than the price of a new cartridge.

Many office supply and stationery stores sell mail-in service kits that allow you to send in empty cartridges for refilling. The mail-in kit costs about 30 percent less than a new cartridge. The kit price includes all shipping costs (both ways) and a special mailing box. Usually, within six or seven days after sending out your empty cartridge, you'll receive the refilled cartridge back in the mail.

For even greater savings, you can buy recharge kits and refill inkjet cartridges yourself. Look for recharge kits at stationery stores and office supply outlets. You may also find them in computer supply catalogs. (See ***Helpful Resources.***)

Printer Ribbons—The nylon fabric ribbons used with dot matrix printers are eminently re-usable. In Japan, workers are required to reink fabric ribbons regularly. But in America, where the throw-away habit is deeply ingrained, about 95 percent of all fabric ribbons are discarded as soon as the print begins to fade.

The practice of reinking is relatively new to America, but it is definitely here. For about $70, anyone can purchase a reinking machine and reink his or her own ribbons. The ink for an average reinking costs about 5 cents. (See ***Helpful Resources.***)

If reinked regularly so it doesn't dry out, a high quality fabric ribbon may be reinked many times. Manufacturers of reinking products suggest that a single ribbon can be safely reinked 10 to 40 times, but some computer maintenance experts advise against reinking more than 10 times.

If a pin from the dot matrix print head gets caught in frayed ribbon fibers, it could be damaged. So when you notice incompletely printed characters, it's probably because the ribbon has begun to fray and it's time for a new one.

If you'd rather not mess with a reinking machine, you can turn faded ribbons over to a cartridge reloader, who will install a fresh ribbon in the old cartridge for about 40 percent less than you'd pay for a new one. A single plastic cartridge can be reloaded several times. Eventually, however, the internal gears wear out and the cartridge is no longer usable. Ask your ribbon dealer to recommend a local reloader, or see ***Helpful Resources.***

Floppy Disks—With proper care, floppy disks can be re-used over and over again. To maximize re-usability, protect floppies from dust, extreme temperatures, moisture, and magnetic fields (televisions, computer monitors, speakers).

When you no longer need the data on a floppy disk, just format the disk. This will essentially erase all data, allowing you to start over with a blank disk.

What happens to packaged software that is still sitting on store shelves when a new version supersedes it? Until a few years ago, the obsolete packages were either sent to landfill, where they take an estimated 450 years to degrade; or they were incinerated, causing acid rain.

Fortunately, this exceedingly wasteful practice is rapidly going out of style. Several environmentally-inspired companies now take the unsold obsolete software disks, remove the labels, erase the software, test and reformat the disks, and repackage them for sale. Although usually known as "recycled disks," technically speaking they are actually re-used. These "recycled" disks are of the highest publisher grade quality and comparable in price to brand name floppies. (See ***Helpful Resources.***)

HELPFUL RESOURCES

1. This company refills inkjet cartridges. For the name of a nearby stationery or office supply store where you can purchase prepaid mail-in kits, contact the company: **GRC, Inc.,** 20650 Prairie St., Chatsworth, CA 91311; (800) 423-5400 or (818) 709-1234.
2. This company sells a motorized reinking machine that supports over 21,000 fabric ribbon cartridges and spool types. **Computer Friends, Inc.,** 14250 NW Science Park Dr., Portland, OR 97229; (800) 547-3303.
3. These companies reload nylon ribbons for dot matrix printers:

    The Ribbon Factory, 2300 E. Patrick Lane, #23, Las Vegas, NV 89119; (800) 275-7422.

    Encore Ribbon, Inc., 1320 Industrial Ave., Suite C, Petaluma, CA 94952; (800) 431-4969 or (707) 762-3544.
4. These companies sell recharge kits that allow you to refill empty inkjet cartridges yourself.

    Computer Friends, Inc., 14250 NW Science Park Dr., Portland, OR 97229; (800) 547-3303.

    American Ribbon Company, 2895 W. Prospect Rd., Fort Lauderdale, FL 33309; (800) 327-1013.

    Global Computer Supplies, 2318 East Del Amo Blvd., Compton, CA 90220; (800) 227-1246.
5. This company manufactures "recycled" floppy disks from unsold obsolete software disks. Their product is available in many office supply catalogs, college bookstores, and office supply stores. Contact the company for the name of a retailer in your area.

    GreenDisk®, 8124 304 Ave., SE, Preston, WA 98050; (800) 305-DISK or (206) 222-7734.

PLASTIC CONTAINERS

Americans go through 4 million single-use plastic bottles every hour! Yet only one bottle in four is recycled; the rest become garbage.

1. Use your purchasing dollars to vote for products that promote re-use and generate less garbage. **Jergens®** brand hand lotion, for example, can be purchased in a refill pouch that generates 78 percent less waste than the regular pump bottle it refills—and costs less too! If the pump wears out (and it will, if you refill the bottle enough times), the manufacturer will send you a replacement for free. Just call the 800 number on the refill pouch. If your store doesn't carry refill pouches, ask them to.
2. Re-use empty yogurt and cottage cheese containers to refrigerate leftovers or to pack lunch foods and snacks. Deli containers can often be re-used too. ***Tip:*** when plastic containers retain food odors, set them outside in a sunny spot for a day and many odors will vanish.
3. If you eat a lot of yogurt, you can reduce plastic container waste by investing in a yogurt maker that utilizes re-usable 8-oz. glass cups. The glass cups come with washable plastic covers so they can be stored in the refrigerator or stashed in a lunch box.

   Homemade yogurt that omits the packaging saves money too. For the price of a yogurt maker (about $20), you can make your own for about half what you'd pay for the store-bought stuff.
4. Some retail nurseries accept plastic pots and flats for re-use. Call first to confirm that the items you have are wanted. And wash containers out before turning them in.

Safety First—Bottles containing nonfood items, including detergents, cleaning supplies, pesticides, motor oil, paint, or any other chemicals, should *never* be re-used to store food or personal care items. If they contained something toxic, don't even rinse them out; follow the manufacturer's instructions for disposal.

PLASTIC FOOD BAGS

Each time you re-use a plastic bread or produce bag, you conserve resources by eliminating the need for a new bag. Since plastic is derived from petroleum, re-using plastic bags conserves oil. Re-using also conserves the energy needed to manufacture and transport the bags.

Obviously, conserving oil slows the rate at which we're depleting the fossil fuel base on which our nation's economy depends. It also minimizes destruction to sensitive habitats caused by oil drilling operations. And by conserving energy, plastic bag re-use diminishes emissions of carbon dioxide, the primary greenhouse gas.

The number of ways plastic food bags can be re-used is limited only by your own ingenuity.

For starters, you can take them back to the grocery store and re-use them to package your produce purchases. Self-serve produce bags make dandy lunch bags.

When packing my suitcase for a trip, I like to place shoes inside a plastic bag first. If I have to pack a wet bathing suit, a plastic bag is invaluable.

Mothers of infants and young children, when away from home, can re-use clean plastic bags to pack a damp washcloth for wiping sticky hands and faces or to stash soiled diapers. (See ***Baby Wipes.***)

Dog owners can use plastic food bags to clean up after their pets. Just put a bag over your hand, scoop, turn the bag inside out, knot the top of the bag, and carry it home for proper disposal. Your neighbors will thank you for your consideration.

Before re-using plastic bags for food, you may wish to wash them. Wet bags can be laid out on the dish rack to dry or you can purchase a wooden rack made for that purpose. These convenient devices can be purchased at environmentally oriented stores and from mail-order catalogs. (See ***Helpful Resources.***)

Instead of plastic wrap or aluminum foil to cover leftovers, use a clean plastic food bag secured with a rubber band.

When plastic food bags are no longer re-usable, you can continue to conserve resources by recycling them. The Plastic Bag Information Clearinghouse toll free hotline maintains a database of over 14,000 supermarkets, retail stores, and dry cleaners across the country where you can recycle plastic grocery sacks, bread bags, self-serve produce bags, newspaper delivery sleeves, dry cleaning bags, and even light colored retail bags. (See ***Helpful Resources.***)

Safety First—***The American Journal of Public Health*** advises re-users to turn plastic bags right side out before using them to store foods. Reversing the bags could put food in contact with printing that may contain toxic lead or cadmium. To avoid contaminating produce and other edibles with bacteria that could cause food-poisoning, bags that have stored meat, poultry, fish, or cheese should not be re-used for food items.

Don't forget to keep plastic bags away from babies and children to avoid danger of suffocation.

HELPFUL RESOURCES

1. Wooden drying racks for plastic bags are available from:

   Seventh Generation, 360 Interlocken Blvd., Suite 300, Broomfield, CO 80021; (800) 456-1177.

   Real Goods, 555 Leslie St., Ukiah, CA 95482-5507; (800) 762-7325.

2. For a plastic bag recycling collection site near you, contact the **Plastic Bag Information Clearinghouse:** 1817 E. Carson St., Pittsburgh, PA 15203; (800) 438-5856.

RAZORS

The Short Sweet Life of a Disposable Razor—Like paper plates and cups, disposable razors are seductively convenient, but problematically short-lived. When a disposable razor loses its edge and you toss it in the trash, it faces an ignominious finish in landfill, where it joins the 180 million other disposable razors that Americans toss out each year. Disposables that are not lucky enough to wind up in landfill will be sending up their last hurrah from an incinerator—into the air we breathe.

A Close Shave—In addition to environmental concerns, there is another good reason to invest in a re-usable razor. Re-usable razors give a closer shave.

According to the panel of men who evaluated shaving systems for *Consumer Reports* magazine in 1995, razors with replaceable cartridges consistently outperformed disposables. The testers found that re-usables gave a closer shave and were easier to use than disposables.

Electric Razors—If you opt for an electric razor, select one with a rechargeable battery that can be removed and replaced when its five-year expected service life is at an end. That way you can have a second battery installed and keep your shaver going for another five years or more. Avoid the less expensive shavers that rely on disposable AA or AAA cells.

HELPFUL RESOURCES

1. **Razor Saver®** is a handy sharpening gadget that promises to revive any cartridge or regular blade (single or double). According to the manufacturer, Razor Saver makes it possible to get 50 or more shaves from any blade—even disposables. Available for about $10 from **Seventh Generation:** 360 Interlocken Blvd., Suite 300, Broomfield, CO 80021; (800) 456-1177.
2. Want to know which razor is likely to work best for you? See the blade razor ratings in the October 1995 issue of ***Consumer Reports*** magazine (often available in the reference section of public libraries).

REFRIGERATORS

There are exceptions to every rule—even the rule that says repairing and re-using is better for the planet than buying a brand new replacement.

New Refrigerators Waste Less Energy—Thanks to stringent new energy-efficiency standards, refrigerators on the showroom floor today use so much less electricity than those manufactured ten to twenty years ago that getting rid of the old box and buying a new one can actually save you money and give the planet a break too.

Buying a new refrigerator that consumes 1000 kilowatt-hours (kWh) less per year than an energy hog manufactured in the 1970s will shave about $100 from your annual utility bill. Over its fifteen-year life expectancy that adds up to $1,500, often more than the original purchase price.

Reducing your refrigerator's annual energy appetite by 1000 kWh benefits the environment too, preventing about a ton of carbon dioxide and twenty pounds of sulfur dioxide from being pumped into the atmosphere every year. (Carbon dioxide is the primary greenhouse gas; sulfur dioxide is the leading cause of acid rain.) By the year 2000, the new appliance energy-efficiency standards are expected to reduce national electricity consumption by 3 percent, offsetting the need for thirty-one large new power plants.

Cut the Cost of Cooling—All of the new refrigerators meet federal energy standards, but some are more energy efficient than others. So it pays to shop around.

Be sure to check out the bright yellow EnergyGuide label that federal law requires on all new refrigerators. This label indicates how efficient an appliance is relative to other comparable models on the market. It also shows how much energy a particular model consumes in a year and translates that energy use into annual operating cost.

For maximum energy savings, ask an appliance dealer about **Whirlpool's Super Efficient Refrigerator,** which exceeds federal energy standards by 30 to 40 percent and is totally free of ozone-destroying CFCs. This environment-friendly fridge is sold under the **Whirlpool, KitchenAid,** and **Kenmore** brand names. Some utility companies offer a handsome rebate to customers who purchase a Super Efficient Refrigerator.

Beware of added features, such as automatic ice makers and through-the-door dispensers. Although manufacturers often suggest that these features save energy because they avoid door openings, they don't.

Automatic ice makers have been shown to increase a refrigerator's energy use by about 20 percent ($24 per year). Through-the-door features such as cold water dispensers and ice dispensers add about 10 percent ($12 per year) to a refrigerator's energy use. Worse yet, according to *Consumer Reports,* refrigerators with ice makers or through-the-door devices are more likely to need repairs.

Maintenance Measures That Pay—Regardless of whether you choose to invest in an efficient new unit or opt to keep your old one chugging along, don't overlook simple energy-enhancing maintenance measures.

Keep the kitchen cool. Because refrigerators are very sensitive to ambient temperatures, the most reliable way to minimize a refrigerator's energy use is to keep the kitchen cool. (This doesn't mean turning on the air conditioning.) If possible, take care not to place the refrigerator next to the stove or where sunlight from a window will warm it and make it work harder.

Make sure there is good air circulation around the fridge. The coils, which are located on the back of the refrigerator or under it, can cool most effectively when there is plenty of air moving across them. Well-ventilated coils have been shown to reduce energy consumption by 15 percent. Storing food and other stuff on top of a fridge can obstruct the flow of air across the coils and run up your energy bill.

Clean the coils. Dirty condenser coils cannot dissipate heat as well as clean ones. When researchers at *Consumer Reports* cleaned the coils on a fourteen-year old fridge, power consumption dropped by 6 percent. Regular coil cleaning (every three months) may also help extend the life of the compressor. Condenser coils are usually located in the back or on the bottom behind the kickplate. Before cleaning them, be sure to turn off the refrigerator at the temperature control.

To clean bottom coils, remove the kickplate and use a narrow long-handled brush or a refrigerator coil brush (available from many appliance stores and some hardware stores) to reach the coils. When you pull the brush out from between the coils, it will be covered with dust. Vacuum the dust from the brush and repeat several times. To clean coils at the back, you may need to pull the refrigerator away from the wall. Be sure to unplug it first, and put something under the front legs to protect the floor. Then pull the fridge (if it has rollers) or gently rock it back and forth to move it out, taking care not to dislodge any water pipes or drip pans. To clean the coils, use a brush or the vacuum cleaner.

When you're done, don't forget to plug the fridge back in and turn it on.

HELPFUL RESOURCES

To obtain product information on **Whirlpool's Super Efficient Refrigerator,** call (800) 253-1301.

SANITARY PROTECTION FOR WOMEN

You've Got to Be Kidding—Of all the different types of re-use I've explored, this is the only one that moves my female friends to roll their eyes and groan: "Re-usable sanitary pads? You've got to be kidding!"

To fully appreciate the beauty of re-usable sanitary pads, one must first understand the environmental devastations associated with the their disposable counterparts.

The Problem with Disposables—Disposable pads are made with a plastic liner (to protect against leakage), which renders them about as nonbiodegradable as the infamous disposable diapers. Each year, American women throw out about 12 billion disposable sanitary pads, most of which end up in landfills where the nonbiodegradable plastic backing remains intact virtually forever.

Both the fluffy white core of disposable pads and the highly absorbent rayon component of tampons are made from wood pulp and are biodegradable. But to give it the appearance of "sanitary" whiteness, the naturally brown wood pulp is bleached with chlorine. This bleaching process synthesizes an arsenal of chlorine-based pollutants known as organochlorines, which are highly resistant to biological breakdown and cause considerable damage wherever they enter an ecosystem. And enter ecosystems they do. Each year, U.S. pulp mills discharge an estimated 400 to 700 million pounds of organochlorines into the nation's waterways.

In spite of printed warnings not to flush plastic tampon applicators, lots of women do it anyway. We know this because beaches and oceans are littered with pink and white "beach whistles" flushed out to sea from cities that still dump raw sewage into the ocean. Sooner or later, the applicators break into pieces, which may be ingested by marine animals mistaking them for food. And when indigestible plastic blocks an animal's digestive tract, it dies. It has been estimated that 2 million seabirds and 100 thousand marine mammals die every year from ingesting or being caught in plastic.

For the women who use them, tampons pose a variety of health risks. All tampon brands have been found to cause temporary vaginal dryness, cell peeling, and even tiny ulcers. Some physicians suspect that tampons may be linked with chronic and recurrent vaginitis. And Toxic Shock Syndrome (TSS), a serious illness that is sometimes fatal, has been definitively linked to tampon use, especially among younger women.

Less Toxic Options—Happily, disposable menstrual products are not our only option. Washable cotton menstrual pads are now widely available in health food stores and by mail from a handful of cottage industries springing up around the country. Neither glamorous nor high tech, washable menstrual pads have a beauty all their own: they empower women with the ability to diminish the environmental devastations caused by disposables. And they represent an elegantly simple step toward our planet's well-being.

Usually made of 100 percent cotton flannelette or cotton fleece, re-usable pads are wonderfully soft and comfortable to wear, and they're available in a variety of designs and absorbencies. Some companies offer hook and loop fasteners or special undergarments to hold pads securely in place. (See ***Helpful Resources.***)

Caring for re-usable pads is simple. Used pads should be soaked in a bucket or bowl of cold water (to prevent permanent staining), and then washed with the regular laundry. Tamara Slayton of the Menstrual Health Foundation in Sebastopol, California, recommends feeding bloody water from soaking pads to house plants or pouring it on your garden. It's very nourishing stuff.

When you are away from the house, used pads can be stored in an airtight plastic bag (a re-used one, of course), then tossed in cold water when you get home.

HELPFUL RESOURCES

1. Washable cotton pads are sold in many health food stores and in environmentally oriented "green" stores. Pads can also be purchased by mail from:

   Feminine Options®, N14397 380th St., Ridgeland, WI 54763; (800) 236-4941 or (715) 455-1875.

   Womankind, P.O. Box 1775, Sebastopol, CA 95473; (707) 522-8662.

   Gladrags, P.O. Box 12751, Portland, OR 97212; (503) 282-0436.

   Allergy Resources, P.O. Box 888, Palmer Lake, CO 80133; (800) USE-FLAX or (719) 488-3630.

2. If you'd like to learn more about the environmental and personal health impacts of disposable menstrual products, I recommend Liz Armstrong and Adrienne Scott's book, ***Whitewash,*** (Harper Collins, 1992, $12.95).

SPORTING GOODS

In 1983, Martha Morris was jobless in Minneapolis and needed rent money. After selling her furniture, she decided to sell her backpack. The problem was, she couldn't find anyone to buy it. When clothing consignment stores refused her backpack, she ran a classified ad, posted notices on bulletin boards . . . all to no avail. That's when she decided to open a store that would buy and sell used sporting goods. "I thought if I needed to sell this, others did too," Ms. Morris recalls.

Armed with $15,000 borrowed from a friend's mother, Martha opened Play It Again Sports. Within five years, she opened two more stores. Fast-forwarding to 1996, there are now over 700 Play It Again Sports franchises in the United States and Canada. Each store markets an array of equipment from fitness machines to boogie boards.

Beat the High Cost of Gearing Up—The folks who shop at Play it Again Sports appreciate the value of preowned sporting goods that are "used—but not used up." Sports enthusiasts on a budget find the items they want at a price they can afford. Upgrading golfers, skiers, and tennis players receive cash for unwanted paraphernalia. And parents save money when they trade in gear that their children have outgrown.

If It's Broken, Get It Fixed—Although general sporting goods stores can repair quite a few items, many of the best repairers are quartered in shops that focus on a specific type of equipment. The camping gear specialists are often experts at fixing broken zippers on sleeping bags, parkas, and tents. Golf shops are equipped to regrip golf clubs. And bicycle shops can, well, repair bicycles.

Rent Instead of Buy—For folks who are experimenting with a new sport, renting makes a lot of sense. If you're organizing a picnic for a large group, you probably won't have to look far for a sporting goods dealer who rents volleyball and softball equipment, badminton racquets and nets, even horseshoes.

Donate Instead of Dump—If your area doesn't happen to have a used sporting goods store, a walk through the Yellow Pages and a few phone calls is usually all it takes to locate a nonprofit organization that's eager to take old gear off your hands. Several national programs actively solicit donations of tennis racquets, golf equipment, and camping gear. Some of these groups are listed under ***Helpful Resources.***

Don't overlook local afterschool recreation programs such as Boys and Girls Clubs of America. You may have just what they need, but you'll never know until you give them a call. In addition, Goodwill, Salvation Army, and other nonprofit thrift shops often accept usable gear. (See Appendix A, ***Donate Instead of Dump.***)

HELPFUL RESOURCES

1. **Inner City Outings** (ICO) gladly accepts all sorts of camping and backpacking equipment in usable or repairable condition. A national special interest group of the **Sierra Club** with many branches in urban areas throughout the country, ICO's goal is to provide outdoor and wilderness experiences for inner city youth. Activities, which vary by venue, include canoeing, hiking, camping, mountain biking, rafting, snow camping, and backpacking. For an ICO near you, call the local Sierra Club chapter or contact the Club's national ICO coordinator at 85 Second St., second floor, San Francisco, CA 94105; (415) 977-5500.
2. **Racquets-for-Kids** collects serviceable old racquets and distributes them to recreation programs and young players who cannot afford to purchase them. Donors get a receipt for the appraised value of their racquet, which they may declare as a charitable deduction on the next tax return. For the name of a participating tennis shop in your area, contact the **Tennis Industry Association** at 200 Castlewood Drive., North Palm Beach, FL 33408; (407) 848-1026.
3. **Clubs for Kids** is a program of the **Professional Golfers Association of America** (PGA) that accepts used clubs and distributes them to kids who can't afford them. For more information, contact the PGA at 1000 Avenue of the Champions, Box 109601, Palm Beach Gardens, FL 33410-9601; (407) 624-8444.

TABLE NAPKINS

Cloth Is Better—It has been estimated that paper napkins produce up to fifteen times more solid waste than their cloth counterparts. If you really want to appreciate how paper napkin waste adds up, consider the case of Red Lobster Restaurants. When this national restaurant chain decided to switch from paper to cloth napkins in 1991, the company estimated that the switch would eliminate 4.7 million pounds of paper waste per year.

Obviously, an individual household uses far fewer napkins than a national restaurant chain, but if enough householders made this simple switch, the environmental impact would be impressive. And since it all has to start somewhere, where better than at your own table?

Colorful cloth napkins can liven up your dining room table. To launder, just toss them in with the rest of the wash. So ironing won't be necessary, be sure to buy napkins made with permanent press fabrics. Cloth napkins add virtually nothing to laundering costs, but they do save trees, preserve precious landfill space, and reduce air and water pollution caused by the paper manufacturing process. (See ***Paper.***)

TABLEWARE

According to the Environmental Protection Agency, Americans toss out nearly one million tons of disposable paper plates and cups every year. When you figure that it takes approximately 17 trees to make a ton of paper, this works out to roughly 17 million trees that we're sacrificing annually for the convenience of disposable tableware.

The solution? Wash dishes. This brilliantly simple, if unexciting, solution is capable of producing genuinely worthwhile results. Fewer paper plates and cups used means more trees saved. And since trees absorb carbon dioxide, this helps combat global warming. Because the paper-making process utilizes immense amounts of energy, and because pulp mills discharge huge quantities of toxic chemicals into streams and coastal waters, fewer paper plates used means less air and water pollution.

Keep a Cup at Work—At your workplace, keep a washable ceramic coffee cup on hand . . . and use it. Encourage your employer to offer a 5 cent discount to those who bring their own ceramic coffee cups.

Simplify Your Life—Avoid single-use plates, cups, knives, and forks like the plague. If your super busy lifestyle leaves you with neither the time nor the energy to wash dishes, this may be an indication that you need to consider simplifying your life and slowing down a bit. You deserve a break, and the planet does too.

Borrow What You Can—When you are hosting a large party but don't have enough plates or silverware, consider borrowing additional dishes and flatware for the occasion from friends and neighbors. You may be pleasantly surprised to discover how eager many people are to help save the planet in small doable ways. You can also rent extra china, silver, and serving pieces from equipment rental outlets that handle party goods. (Look for them in the Yellow Pages under such headings as "Party Supplies and Equipment" or "Rental Service Stores.")

Plastic Is Re-usable—Plastic party cups intended for only one use are actually quite washable. One frugal couple I know attended a neighborhood party, and when helping the hostess to clean up afterwards, they scavenged dozens of discarded plastic wine cups. They brought the cups home, washed them up, and re-used them for their next party. Heavy-duty plastic flatware can also be washed and re-used many times. In fact, at least one manufacturer of plastic flatware actually promotes its product's re-usability by printing the word "re-usable" in bold letters on the front of the package.

Successful Businesses Listen—If you patronize a restaurant that uses disposable plates, ask them to consider switching to washables. Share your concerns with the owner or manager; let them know that you would gladly eat there more often if they could find ways to generate less waste. Successful business people pay attention to what their customers have to say.

TEA BAGS

Paper tea bags are doubtless the most convenient and popular way to prepare tea, but by far the tastiest way to brew this ancient beverage is the old-fashioned way—loose in a teapot.

There are several very practical reasons for eschewing those modern contraptions known as tea bags. Consider first the waste and the wood products that tea bags represent—all those paper bags, wrappers, and tags. You could toss them all on the compost heap, but what about tea bags that come wrapped in nonbiodegradable foil packets? Also worth noting is the fact that tea bags made from paper whitened with chlorine bleach may contain traces of the highly toxic chemical dioxin, which is likely to leach into your drink. (See ***Coffee Filters.***)

If sparing hard-earned dollars is a priority, then listen up: ounce for ounce, loose tea often costs less than half as much as bagged tea. What's more, loose tea is generally of higher quality than the "tea dust" they put in many bags.

If you brew the tea loose in a pot, you can use a special tea strainer (available at most supermarkets) to catch the wet leaves as you pour each cup. Or you can use an infuser (like a tea ball) as a "re-usable tea bag." To allow the tea leaves to expand and release their flavor, fill the infuser only half way. When the tea is as strong as you want it, simply remove the infuser.

Simple stainless steel tea balls are available at most supermarkets for about $2. Or you can order them through the mail (see ***Helpful Resources***). Try specialty shops for infusers in whimsical shapes—miniature tea pots, tiny cottages, etc.

HELPFUL RESOURCES

The ***Stash Tea*** catalog markets a variety of infusers, tea balls, and tea filters. To request a free catalog: 9040 SW Burnham St., Tigard, OR 97223-6199; (800) 826-4218.

TIRES

As a nation, we discard some 240 million tires every year. But as we haven't yet figured out how to re-use, recycle, or dispose of all these tires, we're now encumbered with two to three billion stockpiled tires—enough to circle the globe twelve times!

If tire piles were merely ugly, public officials might find a way to ignore them. But tire piles are worse than unsightly—they pose serious public health hazards. Tire piles provide habitat for rats and breeding grounds for mosquitoes. When tire piles ignite—as several dozen do each year—they belch toxic, foul-smelling smoke, and the remaining oils and soot can pollute surface- and ground-water supplies. Putting out a tire fire and cleaning up the toxic mess it leaves behind costs taxpayers millions of dollars.

Making Old Tires Disappear—Unfortunately, recycling old tires into new ones is not a practical option at this time, largely because vulcanization, the process that makes tires hard and long-lasting, is believed to be irreversible.

Tantalized by a plentiful supply of free raw materials, inventors and entrepreneurs have developed myriad ways to transform discarded tires into useful things: roofing shingles and shoe soles, door mats and mud flaps, rubberized asphalt for highway paving, fuel for cement kilns and electricity generation plants—to name a few.

In spite of this creative ferment, the tire piles continue to grow. According to the Scrap Tire Management Council, Americans discarded 253 million tires in 1995, but only 175 million of them were transformed into useful products or energy, leaving an estimated 78 million with nowhere to go but the nation's growing tire heap.

Making Tires Last Longer—The average passenger tire's life expectancy is 40,000 to 60,000 miles. But a great many tires never see 40,000 miles, simply because drivers fail to maintain them attentively.

It has been estimated that periodic maintenance and thoughtful driving habits could as much as double the number of miles that an average car tire may be safely driven, effectively cutting in half the number of discarded tires. With fewer spent tires to handle, existing technologies could begin to put a significant dent in the growth of tire piles. Here's the strategy:

1. Purchase quality long-lasting tires with a high tread wear grade of 200 or more. Tread wear grades range from 50 to 420. Multiplying the tread wear grade by 200 gives a rough estimate of a tire's expected mileage. For example, a tire graded 100 should last about 20,000 miles; a tire graded 200 should last twice as long.

   It would be a mistake, of course, to expect actual mileage to match tread wear estimates, which are based on the wear rate of a tire when tested under highly controlled conditions. In real life, wear depends on actual road conditions, driving habits, and maintenance practices.
2. Check tire pressure at least once a month. Proper inflation extends tire life, helps prevent accidents, and maximizes gas mileage. Underinflation has been shown to decrease gas mileage by as much as 5 percent. According to the U.S. Department of Energy, if all Americans drove on properly inflated tires, we could save over two million gallons of gasoline every single day of the year.

   Invest $5 in an accurate tire gauge, and inflate tires to the pressure prescribed on the tire placard (usually located on the driver's door, the door post, in the glove box lid, or inside the trunk lid).

   For the most accurate pressure reading, check tires when they are cold, that is, before they have been driven a mile.
3. Unless otherwise directed by the owner's manual, have tires rotated every 6,000 miles or at any sign of uneven wear. Different wheel positions wear at different rates. Rotating the tires evens out the wear, so they'll last longer.

   If tires show uneven wear, ask the service person to inspect for and correct any misalignment or imbalance before rotation. Misalignment will result in faster and uneven tire wear.

4. Cultivate driving habits gentle to tires: avoid fast starts, stops, and turns. Avoid potholes and objects on the road. When parking, don't run over curbs or scrape the tire sidewall against the curb. Avoid tire spinning.
5. Replace tires when the tread is worn down to 1⁄16". When that point is reached, wear bars (narrow strips of smooth rubber across the tread) will appear. By replacing tires when the wear bars appear, you will be driving more safely. You will also increase the tire's potential re-usability, because tires that have been driven to the point of baldness are unsuitable for retreading.

Retread Tires—When it's time to discard worn tires, consider replacing them with retread tires. By re-using the existing tire body and replacing only the tread, retreading conserves significant amounts of oil—about 4 gallons per passenger tire.

Each purchase of a retreaded tire is a substantial act of conservation. In 1994, passenger tire retreaders saved the nation nearly 30 million gallons of oil. When you add in truck, bus, and airplane tires, retreading saves 400 million gallons of oil each year.

Unfortunately, in many areas of the country, retreads for passenger vehicles are not readily available. (Regions subject to heavy snow, where retread snow tires are a popular product, appear to be the exception.)

Fifty years ago, retreads were sold nationwide, but over the last twenty years the number of retreaded passenger tires has fallen off dramatically—from a peak of 35 million in the mid-sixties to 5.3 million in 1995.

Industry experts attribute the decline to low oil prices and the introduction of inexpensive new radial tires, which have caused stiff price competition for retreads. If oil prices rise substantially, however, industry observers look for a reversal of this trend, so stay tuned.

Meanwhile, the good news is that in 1993, President Clinton signed an executive order mandating the use of retreaded tires on all government vehicles. Although some agencies have been slow to comply, in 1994 the U.S Postal Service purchased 77,600 retreaded tires—up 50 percent from two years previous.

In Europe, where limited natural resources and disposal space have exacerbated many environmental concerns, the retreading rate is significantly higher than in the U.S. According to a recent article in ***Scrap Tire News,*** the European Union expects to achieve a 25 percent retreading rate by the year 2000 in the major European countries.

HELPFUL RESOURCES

1. For $4.00 the **Tire Industry Safety Council** will send you an air pressure gauge, a tread depth gauge, four tire valve caps, and a 12-page booklet on tire care and safety. Send check or money order to Tire Industry Safety Council, P.O. Box 3147, Medina, OH 44258.
2. The **Retread Information Bureau** can tell you everything you ever wanted to know about retreads: 900 Weldon Grove, Pacific Grove, CA 93950; (408) 372-1917.
3. You can obtain a free Environmental Fact Sheet on "Purchasing and Maintaining Retread Passenger Tires" by calling the **U.S. Environmental Protection Agency's RCRA Hotline** at (800) 424-9346. Ask for Publication #EPA530-F-95-019.

TOOLS

Whether you're embarking on a home improvement project, changing a tire, or digging a ditch, having the right tool can make it infinitely easier to obtain satisfying results.

Where the Used Tools Are—If you're looking for ways to stretch limited dollars and are willing to spend the time to shop around, good used tools can be every bit as useful as brand new ones. In fact, quality used tools retain their value so well that, during periods of accelerated inflation, some folks have actually toyed with the idea of investing in used tools.

According to the self-described "tool nuts" I interviewed, used tools are among the most popular items offered at yard sales. The *cognoscenti* also hunt used tools at flea markets, swap meets, auctions, and in newspaper want ads. And, although they are few and far between, secondhand tool stores really *do* exist. (Look for them under "Tools" in the Yellow Pages.)

Since most used tools are sold "as is" and carry no warranty, it pays to inspect them carefully before purchasing. If you're not comfortable without a warranty, some factory remanufactured tools are available (usually in tool stores) for about 15 percent less than the cost of new ones.

Looking for Mr. or Ms. Fix-It—When an expensive tool breaks, a tool repair expert can often restore it to full usefulness. Many of the same stores that sell tools also fix them. If not, ask the tool store to recommend a reliable repair service. Or look for repair shops under "Tools" in the local Yellow Pages.

Rent Instead of Buy—If the right tool for the job is an expensive one that you may use only once or twice, it usually makes sense to rent instead of buy. Some hardware stores rent a few small tools, and equipment rental outlets offer almost every conceivable kind of tool. Look for these rental services in the Yellow Pages under "Rental Equipment" or "Tools."

Many rental outlets lend tools for an hour, a day, or a week—whatever you need. Ask agency personnel to demonstrate how to operate equipment properly and safely. (This is also a good way to verify that the equipment is in good operating condition.) Some agencies will even send you home with a demonstration video.

Tool Borrowing Etiquette—Borrowing tools from friends or neighbors is a fine re-use strategy. Just be sure to ask the tool's owner to show you how to use it properly. And remember that if you break it—or even if it simply ceases to function normally while in your custody—it's your responsibility as a borrower to get it repaired promptly.

Tool Lending Library—A few pioneering public libraries operate tool lending libraries. The tool lending library in Berkeley, California houses over 3,000 different hand and power tools. There is absolutely no charge for borrowing, but overdue fines are steep—up to $15 per day.

Donate Instead of Dump—If you've got home maintenance and repair tools to spare, try offering them to the local Habitat for Humanity or to Christmas in April. Or donate them to a tool lending library. (See Appendix A, ***Donate Instead Of Dump***.)

TOOTHBRUSHES

When ordinary toothbrushes are past their prime and no longer suitable for cleaning teeth, they're great for cleaning other things. Nothing beats an old toothbrush for loosening the grunge that collects around tub and sink faucets and between ceramic tiles. Retired toothbrushes are also handy for removing garden dirt from fingernails and cuticles, or scraping mud and dirt from the soles of shoes and sneakers.

Re-usable Toothbrushes?—What will they think of next? These toothbrushes are equipped with removable heads that snap onto a permanent handle. Sort of like popping a fresh blade in your razor.

At present, the chief problem with re-usable toothbrushes is availability. You can purchase them by mail from a few green-oriented catalogs, but you're unlikely to run across them in supermarkets or small drugstores, though I managed to find some at a large chain drugstore.

HELPFUL RESOURCES

Toothbrushes with replaceable heads may be ordered from **Seventh Generation,** 360 Interlocken Blvd., Suite 300, Broomfield, CO 80021-3440; (800) 456-1177.

TOYS

If you are tired of investing a fortune in toys that children rarely remain excited about for more than a few days, you'll be glad to learn that re-use can trim toy budgets radically without sacrificing either quality or quantity. In fact, by tapping into the great reservoir of previously cherished playthings, you're likely to discover you can actually afford more high quality skill-developing toys than ever before.

Handing outgrown toys down to younger siblings, cousins, or children of friends and neighbors is an environmentally sound tradition that conserves the energy and natural resources that would otherwise be required to manufacture new toys. In this way, hand-me-downs effectively reduce air pollution, acid rain, and the devastating environmental damage caused by mining and lumbering operations.

An intriguing new breed of upscale consignment shop that specializes in used children's gear is helping to keep gently used toys out of landfills and in the playful hands of children. As a rule, these shops handle only high quality toys in excellent condition. Some shops pay cash for used toys; others pay in credit toward the purchase of "new" items. The best way to find these stores is in the Yellow Pages. Look for them under headings such as "Toys–Retail" or "Children's and Infants' Garments-Retail."

Shoppers who know precisely what they're looking for can find some real bargains in the classified ads. By simply placing a want ad in the local newspaper, Boston Computer Exchange President Alex Randall obtained exactly what his three-year old son wanted: "Enough **Lego®** sets to build a small city." For about $300, Randall purchased Lego sets that, if new, would have cost between $3,000 and $4,000.

Some of the very best bargains in used toys surface at garage and rummage sales.

When previously cherished playthings need a bit of sprucing up, a touch of elbow grease can often do the trick. For example:

- Many plastic and wooden toys can be easily cleaned with warm soapy water and an old toothbrush. If you're concerned about the presence of infectious organisms, the manufacturers of **Clorox®** recommend adding 3/4 cup regular household bleach (5.25% sodium hypochlorite) to a gallon of warm soapy water. Soak toys in solution for 2 to 5 minutes, then rinse thoroughly and allow to air dry. Leaving toys outside in the sun for a day is also a good idea, as ultraviolet light kills many bacteria.
- Broken toys can often be repaired by ordering replacement parts from the manufacturer. You will need to know the company name as well as the toy's name and model number. To find out if the manufacturer has a toll-free customer service hotline, call the phone company at (800) 555-1212. (See ***Helpful Resources.***)
- Occasionally, two broken toys can be combined to make a new one. Look for a second toy that you can dismantle for parts at garage sales, thrift stores, and flea markets.
- Tricycles, bikes, and other children's vehicles can be refreshed with a coat of rust-inhibiting paint. Before painting, use steel wool or a wire brush to remove loose rust. Avoid using leftover paint that might have been purchased before 1978, when the Consumer Product Safety Commission virtually banned lead in new paint. (See ***Safety First!*** below.)
- Revitalize dog-eared game boxes with a covering of colorful wrapping paper.

Find out if there is a toy library in your town. Most toy libraries are nonprofit volunteer-run organizations where kids can check out safe high quality toys for free or a nominal charge. (See ***Helpful Resources.***)

Even if you choose to purchase brand new toys for your child, you can still participate in the re-use process by donating outgrown items to a local charity. Goodwill Industries, Salvation Army, and many other thrift shops are delighted to receive toys in good condition. Many toy libraries also welcome donations of durable high quality items. (See ***Helpful Resources.***)

The November-December holiday season presents an excellent opportunity to clean out children's closets in order to make room for gifts that will be arriving soon. Enlist children's assistance in determining which toys are no longer wanted. Then plan a family outing around delivering the selected items to a local charity. This is a good way to teach children the real meaning of the holidays and introduce them to the pleasures of giving.

Safety First!—Each year thousands of children suffer toy-related injuries. Whether shopping at garage sales or department stores, parents must be alert for potentially unsafe toys. Here are just a few of the hazards to be on the lookout for:

1. Choking hazards and swallowable parts (especially for children of the oral age—up to five years);
2. Strangulation risks (strings longer than 7 inches);
3. Flammable materials;
4. Toys that utilize household current that could shock or burn;
5. Sharp or rigid points that could blind a child (especially projectile weaponry);
6. Toxic chemicals in chemistry sets or cosmetics kits. Avoid activity kits that don't have complete instructions;
7. Stuffed animal seams that could separate, allowing the child to ingest the stuffing;
8. Suffocation hazards;
9. Age recommendations are usually on toy packaging. If you're buying a previously cherished toy without packaging, make sure the toy you're buying is age-appropriate.

Maintenance Safety Tip—Parents should inspect children's toys regularly for broken parts that might pose a choking hazard for small children. Vehicles such as bicycles and tricycles should be checked for signs of rust or damage that could prove hazardous.

HELPFUL RESOURCES

1. **Fisher-Price** publishes a "Bits & Pieces" catalog of replacement parts for the toys they manufacture. For a free copy, contact **Fisher-Price Consumer Affairs,** 636 Girard Ave., East Aurora, NY 14052-1880; (800) 432-5437.
2. To find out if there is a toy library in your area, contact **USA Toy Library Association,** 2530 Crawford Ave., Suite 111, Evanston, IL 60201; (708) 864-3330.
3. The **Los Angeles County Department of Public Social Services** operates a Toy Loan service that allows children to borrow toys from thirty-two Toy Loan Centers just as they borrow books from the public library. Citizens donate thousands of discarded toys that are repaired, painted, and then distributed to toy lending centers. For more information about this program, contact the **Toy Loan Coordinator** at 2200 Humboldt St., Los Angeles, CA 90031; (213) 226-6286.
4. An hour spent reading Edward Swartz' ***Toys That Kill*** (Vintage, 1986) will raise a parent's toy hazard awareness and could save a life. This book is out of print now, but I found several copies at the public library.
5. The **U.S. Consumer Product Safety Commission** publishes several instructive brochures on toy safety and a list of current toy recalls, including:

   Recalls & Corrective Actions for Toys and Children's Products
   Toy Safety: Tips for Consumers
   For Kids Sake, Think Toy Safety

   For a free copy of any of these publications, write to: **U.S Consumer Product Safety Commission, Publication Request, Office of Information and Public Affairs,** Washington, D.C. 20207. If you have questions regarding specific toys, you may call the Commission at (800) 638-2772.

TRASH BAGS

The best packaging for garbage is garbage. Instead of wasting money on plastic trash bags that will only be thrown out, package your trash in bags that have already been used for some other purpose.

The garbage receptacle in my kitchen has been lined with a variety of previously used containers:

- Plastic grocery bags work like a charm. Loop the handles over each end of the kitchen wastebasket, and the fit is just about perfect.
- Paper grocery bags are OK too, unless there's a lot of wet stuff in the garbage, in which case the paper sack may leak or the bottom collapse.
- Dog food sacks, which are often lined with plastic, make first rate garbage bags. Ditto for other types of large sacks that originally contained rice, cat food, cat litter, etc.
- On occasion, I've dropped a large detergent box into the wastebasket. It doesn't hold as much as a grocery bag, but it sure does the job. If a smaller container means taking the trash out more often, that's OK too. Frequent emptying helps minimize household odors.

Better yet, purchase detergent in flexible plastic refill bags and use the empty bags for trash disposal. Plastic refill bags for **Tide®** and **Cheer®** brand detergents utilize 80 percent less material than the paperboard boxes they're designed to refill—and they contain 25 percent recycled plastic.

TROPHIES

When it's time to part with treasured old trophies, don't send your laurels to landfill. Instead, win yourself a "conservation award" by offering them to a local recreation center or high school.

It's easy and inexpensive to create an impressive "new" trophy by removing the old name plaque and replacing it with a new one. On handsome trophies made of metal and wood, even the figurines on top can be changed, allowing a dancer's prize to be re-awarded to a tennis player.

Installing a new plaque on a less expensive plastic trophy may not save money (since a new plaque can cost as much as a new plastic trophy), but it can conserve resources and extend the life of local landfill.

VACUUM CLEANER BAGS

When I was a girl, my mother emptied her re-usable cloth vacuum bag onto an opened newspaper, then bundled the debris into a tidy compact package that she tossed in the garbage can. Compared to the convenience of today's disposable bags, mother's cloth bag emptying was a dusty, messy process. It was also extremely economical.

If you don't mind the mess and if economy is a high priority, you'll be glad to know that re-usable cloth bags are still available for some modern vacuum cleaners. You probably won't find cloth bags at the supermarket, but plenty of vacuum dealers stock them or can order them for you. A re-usable cloth bag that will last for many years costs approximately $20—about what you might pay for 20 inexpensive disposables.

If you're in the market for a new vacuum, be sure to ask which models accept cloth bags. Most commercial models rely on cloth bags, but only some household models accommodate them. **Sanitaire** and **Eureka** continue to manufacture household models designed to take cloth bags.

Most of the half dozen vacuum dealers I interviewed on this subject praised disposable bags over cloth. Only one outspoken dealer suggested that the reason you won't see cloth bags displayed conspicuously in most shops is they are not a profitable item for the dealer.

The problem with cloth, as it was explained to me more than once, is that it doesn't filter out super-small dust particles as effectively as high quality disposables do. As a result, a certain amount of fine dust is sucked out the back of the bag, across the motor, and exhausted out into the air you breathe. (Allergy sufferers take note.) As dust accumulates on the motor, more frequent servicing may be required. For the cost of servicing or repair, one can buy quite a few disposable bags. Not to mention the inconvenience of schlepping your vacuum to the repair shop. This is one of those times when each householder must weigh the advantages of re-use against the disadvantages. If saving money on bags is a high priority, then cloth bags may be the way to go. If dusty vacuum exhaust is a problem, you may be better off with disposables designed to filter small dust particles.

VIDEO GAMES

In 1988, David Pomije had a hunch. He suspected that somewhere among America's 75 million thumb pad warriors there lurked a vast and untapped market for previously played video games. Mr. Pomije set out to test his hunch by renting **Nintendo®** games to video stores and selling them by mail from his home. The response was overwhelming. He named his tiny company **FuncoLand**, and within seven years it grew to include 182 used video game stores, a lively mail-order exchange for used games, and a magazine with 110,000 paid subscribers.

At FuncoLand stores customers can choose from approximately 2,000 different titles, including hard-to-find games that are no longer manufactured. Each store is equipped with video game sampling areas, where customers can try out games before buying them. Most titles come with a 90-day replacement warranty.

Customers who want to sell the video games they no longer play may choose between receiving a check or store credit toward future game purchases. A list of prices the company will pay for each game is posted in each store and updated periodically.

FuncoLand doesn't have a monopoly on previously played video merchandise. In addition to several smaller chains that buy and sell used games, there are plenty of independent used video dealers in cities across the nation. Also, a growing number of retailers that once dealt exclusively in new games have added used items to their inventory. In the town where I live, the used toy store recently added video games to its stock of preowned stuff. If there isn't a used video exchange in your area, you can buy and sell used video games by mail. (See ***Helpful Resources***.) And if you prefer to rent instead of buy, you'll be glad to know that many of the same outlets that rent movies for home VCR viewing also rent video games—ideal if you want to try a game out before deciding to buy it. Some outlets will even rent the equipment to play them on.

Maintenance Tip—To keep video game systems in top working condition, regular cleaning is essential. When units are not cleaned, residues accumulate on electrical contacts in game units and cartridges, causing poor picture and sound and playing information. FuncoLand is so convinced that cleaning extends product life that the company offers customers who purchase a cleaning kit a full one-year replacement warranty on most products.

HELPFUL RESOURCES

For used video outlets in your area, consult the local Yellow Pages. To locate mail-order outlets for used games, I recommend perusing some of the video game magazines that are available at large bookstores and newsstands.

These mail-order dealers will send you a free sales catalog upon request:

Funco, Inc., 10120 West 76th St., Eden Prairie, MN 55344; (612) 946-8100.

Video Game Network, 6233 University Ave., NE, Fridley, MN 55432; (612) 574-9882.

WASHING MACHINES

Let's face it, most of us are too darned busy for our own good. We're overworked and overscheduled. And to sustain this pace, we've become abjectly dependent upon a stable of time-saving household appliances.

When a busy person's trusty washing machine quits working, something must be done about it—and fast. Being chronically short on time myself, I don't have much patience with dysfunctional appliances. On the morning that my own nine-year old washer broke down, getting that machine up and running was such a high priority that I took a half day off from work to stay home and wait for the repairman.

While waiting for the repairman, I speculated gloomily, what if this was just the first of many repairs to come? Perhaps repairing would be a waste of money. Maybe I should bite the bullet and buy a new machine.

It is reassuring to know that plenty of other washing machine owners share my anxieties. About 25 percent of those who responded to *Consumer Reports'* Annual Questionnaire in 1992 chose the trash heap over the repairman in a similar situation.

When the repairman finally arrived and examined my paralyzed appliance, he diagnosed the problem as a worn out belt drive. He assured me that my despair was probably premature; that even though the belt drive had lasted only nine years, the motor could easily endure for twenty. Today, six years later, my washing machine still works fine.

Deciding to repair turned out to be good for my pocketbook. It was also good for the planet. That's because repairing a complex durable product not only conserves its functional value, it conserves the energy and natural resources that would be needed to manufacture a new one. Less energy used means less air pollution generated. And fewer natural resources consumed—in this case, metals—means less environmental damage of the ugly (and often toxic) sort caused by modern mining operations.

Restoration benefits local economies too. Businesses that repair, recondition, reupholster, and rebuild products provide jobs for local skilled workers.

Next time you find yourself wondering whether to repair or not to repair, remember that restorative measures like repairing often have a healing impact on our environment, the local economy, and individual pocketbooks.

A couple of years ago, I ran across a fascinating article in the now-defunct *Newsletter for the National Association of Dumpster Divers and Urban Miners.* The article related the story of how, when her washing machine quit, Rosemary Fuller-Thomas reluctantly made plans to buy a new one. But Rosemary's husband, out of insuppressible curiosity, took it apart and discovered that a tiny twig lodged inside the impeller had caused the water pump to seize up. When he removed the twig, the dead machine came back to life.

Guessing that plenty of other people had probably thrown out perfectly good washing machines because of equally minor problems, Rosemary and her husband began hauling home discarded washers and dryers. They fixed any minor problems they found, ran a few test loads of laundry and sold them—with a 30-day warranty—through want ads in the newspaper.

That is how it started. Today, Rosemary's family operates The Washing Well Store in Portsmouth, Virginia, where they recondition washers and dryers and resell them with a 90-day warranty. They sell used parts (for less than half the cost of new parts) to home handypersons who want to fix their own machines. They also do in-home repairs and offer clients a choice of new or used parts. Rosemary is extremely proud of the fact that in their community's depressed economy, her family has established a thriving new business that saves money for customers at the same time it keeps tons of used appliances out of the waste stream. (Also see: ***Appliances.***)

HELPFUL RESOURCES

When shopping for a washing machine, whether new or preowned, it's a good idea to check out the latest evaluation in ***Consumer Reports*** magazine. This evaluation usually includes a "Frequency of Repair" record for major washing machine manufacturers. The washing machines that require the fewest repairs may initially cost more than those with less impressive track records, but the extra expense may be worth it. Before dismissing the higher-priced options, don't forget to factor in the cost and inconvenience of potential repairs. Many public libraries keep back issues of *Consumer Reports* in their reference collection.

WINDSHIELD WIPERS

When the rains finally came to California after seven years of draught, the windshield wipers on my car could not cope. They had lost their flexibility and grown hard with age; they needed replacing.

I drove to the auto parts store, where I asked a salesperson to help me find a new rubber blade of the proper size. She took one look at the wiper assembly in my hand and said, "Sorry, that's disposable and doesn't take refills. You can't replace just the squeegee. You'll have to buy a complete assembly."

This was news to me. It turns out that a whole lot of windshield wipers are single-use tossables. So if your car has disposable wipers, next time you're in the market for new ones I recommend investing in a refillable assembly. That way, in the future, you'll need to purchase only the blades.

Extending Squeegee Life—Accumulated dirt and grime can impair wiper performance. When wipers smear rather than clear the windshield or when they leave behind a swath of water beads, try cleaning the windshield and wiping the squeegees with a solution of vinegar and water. If that doesn't help, the squeegee probably needs replacing.

WIRE CLOTHES HANGERS

For some time now, I've been quietly conducting an informal poll in my community to determine how many dry cleaners take back wire hangers for re-use. So far, I've found that practically all cleaners gratefully accept hangers for re-use, though only one or two actively solicit them from customers. So it was refreshing to learn that, in contrast, dry cleaners in New York City openly flaunt hanger re-use and energetically encourage patrons to return them.

Concerned that residents were sending 25 million pounds of used clothes hangers each year to the city's rapidly depleting landfill, the New York City Department of Sanitation teamed up with the Neighborhood Cleaners Association to produce a poster urging customers to bring in hangers for re-use.

In addition to keeping hangers out of the waste stream, re-using cuts the dry cleaners' costs. For the Chris French Cleaners in the East Village, customer returns have reduced the amount of hangers it orders by 120 cases per year. This saves the company about $3,600 annually and keeps almost 5,000 pounds of hangers out of the city's overburdened waste stream.

HELPFUL RESOURCES

If you'd like to learn more about the New York City Department of Sanitation's joint program with the Neighborhood Cleaners Association, contact the **Bureau of Waste Prevention, Reuse and Recycling Dept.,** 44 Beaver St., 6th Floor, New York, NY 10004; (212) 837-8183.

ZIPPERS

Just because the zipper in a favorite garment fails doesn't mean you have to pack the garment off to a garage sale or shove it to the back of the closet indefinitely. It means you have to find a good dry cleaner. My dry cleaner will install a new zipper in a pair of slacks or a skirt for $7.

Similarly, when a sleeping bag zipper needs replacing, contact an outdoor equipment store. If they aren't equipped to fix it, they can certainly tell you where to take it for repairs. Broken zippers on purses or luggage can sometimes be revived or replaced at a shoe repair shop or luggage store.

Savvy sewers who know how to replace a broken zipper have a valuable skill. One yard sale buff I know swears that some of the best clothes she has ever stumbled on at neighborhood sales are in fine condition, except for the zipper. She acquires them for peanuts, takes them home, and sews in a new zipper. Actually, many of the zippers she installs are not new at all. They're previously used, salvaged from worn-out old garments on their way to the rag bag.

Zipper First Aid—When zipper teeth or coils break, replacement is usually the only solution, but a few zipper problems are remediable.

1. When metal zippers are stuck, try lubricating them back into action by rubbing the metal teeth with beeswax or a wax candle stub.
2. Old jeans with runaway zippers that refuse to stay up can be controlled with a special safety latch that hooks onto the zipper pull tab (the little metal piece that you grip when you zip). After zipping up, a C-hook slips around the button so the zipper stays closed. (See ***Helpful Resources.***)
3. In a pinch, you can replace a missing pull tab with a small safety pin or a paper clip, provided there is a hole in the slider where the pull tab was originally attached.

4. Broken pull tabs on expensive zippered items such as purses or sleeping bags can often be fixed with a **Zipper Pull Repair Kit®**, which provides the hardware and instructions needed to replace the broken pull tab. (See ***Helpful Resources.***)

Maintenance Tip—Save wear and tear on bottom stops by closing up zippers before laundering.

HELPFUL RESOURCES

1. **Zipper Pull Repair Kits®**, and C-hooks for jeans with runaway zippers may be ordered from **Clotilde's:** 2 Sew Smart Way B8031, Stevens Point, WI 54481-8031; (800) 772-2891.
2. Look for zipper installation instructions in any basic sewing book, such as ***Singer® Sewing Essentials*** (Cy Decosse, 1984, $14.95).

Appendix A
DONATE INSTEAD OF DUMP

Donating usable goods not only helps reduce solid waste, it benefits your community and allows you to feel good about helping others. As an extra bonus, donating can yield a nice tax deduction.

With certain limits, individual taxpayers can claim the fair market value of donated goods as a charitable contribution. Be sure to keep good records of how you determined the fair market value (classified ad clippings, current price guides, visits to local thrift shops, discussions with dealers, etc.). And don't forget to obtain a receipt from the charity describing the nature of the transaction and donated goods. If the value of a donated item exceeds $5,000, you must obtain a written appraisal from a qualified appraiser.

In order to qualify for a tax deduction, the recipient of your gift must be a public charity, public school, public park and recreation program, church, or other nonprofit organization. To make sure the institution receiving your goods is recognized by the IRS as a nonprofit organization, ask for a copy of the organization's exemption letter, or contact the Internal Revenue Service (800-829-1040) and ask if the group is exempt.

Charitable donations made by businesses may also be tax deductible. Although assets that have already been fully depreciated may not be deducted, all related shipping and transportation costs are generally deductible. In order to determine the deductibility of your donation, your tax preparer will need to know:

- How much you initially paid for donated items;
- How much of the purchase cost has been written off through depreciation;
- The current fair market value of the donated goods.

HELPFUL RESOURCES

The **U.S. Internal Revenue Service** produces two free publications that explain how to determine the fair market value of donated property and the records you'll need to keep. These documents may be obtained by phoning the IRS at (800) 829-3676.

Publication 526, ***Charitable Contributions,*** addresses donations valued at less than $5,000.

Publication 561, ***Determining the Value of Donated Property,*** focuses on big-ticket items worth more than $5,000.

Appendix B

ORGANIZATIONS

Global Action Plan
(914) 679-4830
P.O. Box 428
Woodstock, NY 12498

Brings together small groups of family members, friends, and neighbors into "Household Ecoteams" who support each other in voluntarily adopting earth-friendly consumption habits. The goal is to develop a "critical mass" of environmentally responsible individuals whose sustainable lifestyle choices will gradually spread to mainstream culture.

Institute for Local Self-Reliance
(202) 232-4108
2425 18th St., N.W.
Washington, DC 20009-2096

A think tank that has conducted extensive studies of organizations that redistribute secondhand durable goods. Studies focus on how successful re-use organizations benefit the community; how such organizations are structured, managed, and financed.

Materials for the Future Foundation
(415) 561-6530
P.O. Box 29091
San Francisco, CA 94129-0091

Provides grants, loans, business advice, and policy advocacy for community-based re-use and recycling enterprises.

National Association of Resale & Thrift Shops
(800) 544-0751
20331 Mack Avenue
Grosse Point Woods, MI 48236

A trade organization representing the resale industry. Members receive a monthly newsletter and access to books and tapes, workshops, and an annual conference on operating a successful consignment shop.

Reuse and Repair Council
c/o California Resource Recovery Association
(916) 652-4450
4395 Gold Trail Way
Loomis, CA 95650

This California-based group works to encourage legislation that promotes re-use; to educate the public about the important role that re-use and repair play in achieving waste reduction goals; to provide policy direction to local government officials and agencies.

Reuse Development Organization (ReDO)
Contact: Mary Lou Van Deventer at Urban Ore, Inc.
(510) 232-7724
6082 Ralston Ave.
Richmond, CA 94805

The mission of this recently formed group is to promote re-use as an environmentally sound, socially beneficial and economical means for handling unwanted and discarded materials. ReDO plans to provide organizations that sell or distribute secondhand goods with educational materials and a network of peer support.

Used Building Materials Association
(204) 947-0848
#2–70 Albert St.
Winnipeg, Manitoba
Canada R3B 1E7

This brand new organization plans to lobby for legislation conducive to re-use in the U.S. and Canada. It will develop reports and manuals on practical aspects of operating a successful used building materials store.

Wastewise
U.S. Environmental Protection Agency
(800) EPA-WISE
401 "M" St., SW
Washington, DC 20460

Operated by the U.S. Environmental Protection Agency, this program assists businesses in cutting costs by preventing waste. Program recommendations include many types of re-use.

Appendix C

PUBLICATIONS

Discarding the Throwaway Society by John E. Young (Worldwatch Institute, 1991, $5). In a concise thirty-five pages, the author examines the environmental mess we're in and frames a strategy aimed at evolving a less planet-damaging society. Needless to say, re-use plays a leading role in Mr. Young's plan to save the planet.

Journey for the Planet (Global Action Plan, 1995, $15.95). A five-week adventure designed to empower 9 to 12 year-old children to translate their concern for the Earth into eco-friendly consumption choices. The book may be purchased directly from the publisher, **Global Action Plan,** at P.O. Box 428, Woodstock, NY 12498; (914) 679-4830.

How Much is Enough? The Consumer Society and the Future of the Earth by Alan Durning (W.W. Norton, 1992, $8.95). Readers interested in protecting the Earth's future by curtailing their own consumption will find plenty of support in this short, lively book. The author encourages readers to curb material waste by cultivating what he calls "true materialism"—i.e., taking good care of material things by maintaining, mending, and repairing them.

Challenging the consumerist notion that acquiring material possessions brings happiness, Mr. Durning suggests that greater satisfaction may be found in nonmaterial sources of fulfillment such as family and social relationships, meaningful work, and leisure.

How to Restore and Repair Practically Everything by Lorraine Johnson (Viking-Penguin, 1990, $17). Experts in leather, textiles, metal, wood and eight other materials provide step-by-step instructions for dozens of cleaning and restoring processes.

The Re-User. This quarterly newsletter is the voice of the recently formed Used Building Materials Association (UBMA), an organization of for-profit and nonprofit companies and organizations in Canada and the United States that acquire and sell used building materials. To receive a sample copy, contact the UBMA at #2-70 Albert St., Winnipeg, Manitoba, Canada R3B 1E7; (204) 947-0848.

Simple Living: The Journal of Voluntary Simplicity. This quarterly journal serves as a forum for people interested in trading the rat race for a more serene, economically sane, and environmentally sound way of life. Readers share wisdom and experience. Sample copies are $3.75. Contact the publisher at 2319 N. 45th St., Box 149, Seattle, WA 98103.

The Tightwad Gazette, Vol. 1 and 2 by Amy Dacyczyn (Random House, 1993, $9.95). Upbeat and down-to-earth, these volumes offer many bright and environmentally sensitive ideas for saving precious dollars, time, and natural resources.

The ULS Report: Helping People Conserve Resources and Reduce Waste by Using Less Stuff. Addresses ways householders can conserve natural resources by, well, using less stuff. This bi-monthly newsletter is free for the asking. Contact the publisher, **Partners for Environmental Progress,** P.O. Box 130116, Ann Arbor, MI 48113; (313) 668-1690.

Voluntary Simplicity: Toward a Way of Life That Is Outwardly Simple, Inwardly Rich by Duane Elgin (William Morrow, 1993, $10.00). Originally published in 1981, this visionary book explores sustainable ways of living on the planet: frugal consumption, ecological awareness, and personal growth.

ABOUT THE AUTHOR

Kathy Stein has been an enthusiastic environmental activist since the early 1970s. In the 1980s, she led a successful campaign to bring curbside recycling to her hometown in northern California. She now enjoys sharing the good news about re-use with others. She lectures on re-use to college classes as well as professional and community organizations. Stein writes: "I believe that re-use is more than a waste-reduction strategy. It is an ethic, a way of life that demonstrates a deep respect for the Earth and the materials we take from it."

Stein lives in Kensington, California, with her husband John.